£3 8/25

GW01458541

To Geoff

Thank you for your
wise & kind support.
Good luck for the
future

Raj.

Hoverflies
of Surrey

Hoverflies
of Surrey

ROGER K. A. MORRIS

SURREY WILDLIFE TRUST

Cover illustration: Episyrphus batteatus, by David Element

ISBN 0 9526065 3 4

British Library Cataloguing-in-Publication Data.
A catalogue record for this book is available
from the British Library.

First published 1998
by Surrey Wildlife Trust
School Lane, Pirbright, Woking, Surrey GU24 0JN.

Printed and bound in Great Britain by
Biddles Ltd, Guildford and King's Lynn

FOREWORD

Surrey is a very special place because of its wealth of wildlife and the fact that so many wild places are accessible to the public. It is the place where I gained so much of my early entomological experience, leading me to the revelation that some single sites in the county may support at least half the British hoverfly fauna.

Among the most conspicuous insects are the hoverflies. They are among the few insects with a widespread public appeal as being friendly. It was thus that I was persuaded to write a complete identification guide, *British Hoverflies*. In the preface, I expressed the hope that the book would stimulate increased recording and study. That Surrey should now have its own county atlas to hoverflies is a wonderful step forward.

The dedication of Roger Morris and his small team of helpers has led to an excellent county review. It is also important nationally for the insight into factors affecting distribution patterns. Because hoverflies have such varied biologies and ecology, this atlas indicates much about site quality, with important implications for conservation strategy in Surrey.

However, Roger would be the first to admit that the present atlas is no more than a step towards improving knowledge of Surrey hoverflies. This atlas will hopefully act as a stimulus to others to record hoverflies and fill in the gaps in recording, and to improve understanding of the ecology of individual species. The countryside and climate are ever changing in subtle ways and we need to monitor the response and well-being of the fauna. In that sense, it is everyone's responsibility to ensure that Surrey remains a rich county for wildlife, including hoverflies.

ALAN STUBBS

CONTENTS

Preface.. 1

Introduction

Surrey – the study area 3

Geology and biogeography 5

Hoverfly biology and form 10

Habitat associations... 12

Some important sites in Surrey 23

Recording hoverflies in Surrey................................ 28

Data collection, storage and retrieval 32

Recording techniques 34

Biological observations 36

Comparison with other recording schemes 37

Hoverfly conservation 38

Hoverflies and biodiversity..................................... 40

Acknowledgements 42

Explanation of species accounts 44

Species accounts

Bacchini 51

Paragini 64

Syrphini.. 65

Callicerini 108

Cheilosiini 109

Chrysogastrini 131

Eristalini 147

Merodontini... 159

Pelecocerini... 162

Pipizini 163

Sericomyiini.. 174

Volucellini 175

Xylotini 179

Microdontinae... 192

Appendices

1. Checklist of Surrey hoverflies 197
2. Flowers visited by hoverflies 202
3. Index of plants listed by common name 217
4. Species list for Great Bookham Common............... 223
5. Organisations 225
6. Bibliography.. 226
7. Glossary 239

Index 242

PREFACE

The British fly fauna comprises over 6,700 species and includes many families which are difficult to identify for want of readily accessible literature. The hoverflies (Syrphidae) are one of the better studied families with over 270 species listed. Their popularity mainly arises from the publication of *British Hoverflies* by Alan Stubbs and Steven Falk in 1983, a book which has spawned a plethora of county recording schemes, considerable activity in the national recording scheme and an interest in flies that would have previously been unthinkable. This atlas of Surrey hoverflies is itself part of that legacy and has been researched since 1985 when my former interest in moths was replaced with an enthusiasm for hoverflies.

Surrey is an unusual county because it retains a considerable area of semi-natural habitat, a factor which should be borne in mind when commenting upon the frequency of species which are listed as Red Data Book and Nationally Scarce in Falk (1991). The surface geology comprises a rich mosaic of sands and gravels, chalk and clays, and is responsible for much of the variation in distribution in the more habitat-specific hoverflies. The county is, however, effectively land-locked with access to the tidal River Thames only within suburban London. This means that the hoverfly fauna is mainly restricted to non-maritime species and makes the county list of 209 species, known to have been resident, all the more remarkable.

This is the fourth in the series of county atlases covering various invertebrate groups in Surrey. It arose as part of an initiative by members of the Croydon Natural History and Scientific Society to map the insect fauna of Surrey, and has been written to accommodate the needs of both the hoverfly specialist and also those with a passing interest in hoverflies or the wildlife of Surrey. Consequently, there are elements of the book which might be gained just as easily from other texts, but it is hoped that these will help to explain the biology of the Surrey fauna if read, for example, by a nature reserve manager who might not have access to more specialised texts.

The distribution maps in this book have been produced by the program DMap (in its Windows version) written by Dr Alan Morton of Imperial College at Silwood Park. For further information about the program and its implementation in this case, see Morton and Collins (1992).

No atlas project can be successful without contributions from a large number of people, all of whom I hope are listed in the acknowledgements section. In addition to these much-valued contributors, I would particularly like to thank Dr Francis Gilbert and Dr Graham Rotheray for refereeing the text; Dr Stuart Ball, Steven Falk and Alan Stubbs for their helpful comments on earlier versions; Graham Collins for correcting my grammar and checking for inconsistencies, and for creating the map files for inclusion in this book;

Roger Hawkins for proof-reading the final product; Lajla White for preparing the diagrams on page 11; and Denise Ramsay for compiling the glossary and offering helpful suggestions on ways of presenting technical terms. Special thanks must go to Clare Windsor, Peter Abbott and Paul Wickham of Surrey Wildlife Trust who have helped in the production of this book. I am also most grateful to Dr Stuart Ball, Graham Collins and David Element whose magnificent slides have made it possible to produce the colour plates reproduced here, and to Peter Chandler for allowing me to use the names from his new Checklist in this volume. Finally, I would like to thank Dr Lawrence Jones-Walters, my Team Manager, for allowing me study leave to complete this text when time was running out and I was at my wits' end to finish it.

ROGER MORRIS

SURREY – THE STUDY AREA

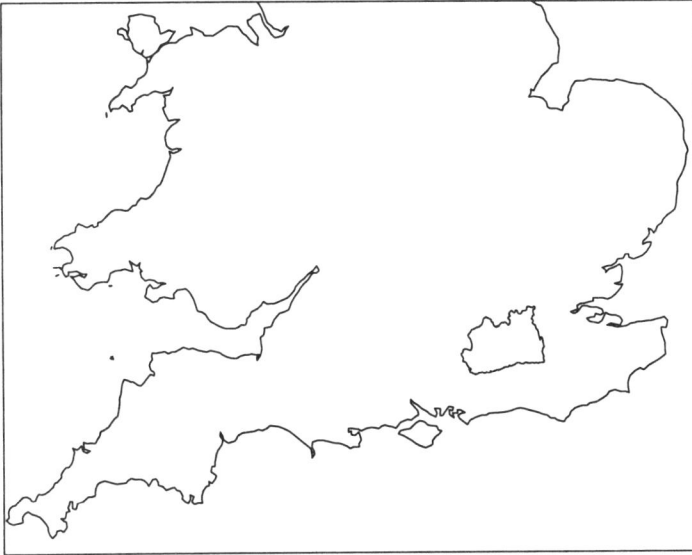

Surrey in relation to southern England

The current county of Surrey lies within a political boundary which has changed markedly in recent times and will continue to remain unstable as political whims and thinking change. This instability makes such boundaries unsuitable for biological recording – a problem which was addressed by H.C. Watson who, in 1852, proposed a series of vice-counties of roughly equal size for botanical recording. His system, which is explained by Dandy (1969), was readily adopted by botanists and has since been adopted by zoologists; it is still in use today.

The differences between the vice-county of Surrey and the modern county are most apparent on the northern boundaries where the vice-county follows the River Thames and includes the south-western quadrant of Greater London; it excludes the district of Spelthorne which only became attached to Surrey in 1965 and in fact belongs principally to the vice-county of Middlesex. The southern boundary differs around Horley where it embraces Gatwick airport, which currently lies in West Sussex, and on the western boundary of present-day Surrey an area of approximately one square kilometre to the south of the village of Batt's Corner is excluded. The vice-county boundary and those of its immediate neighbours are shown on the map overleaf.

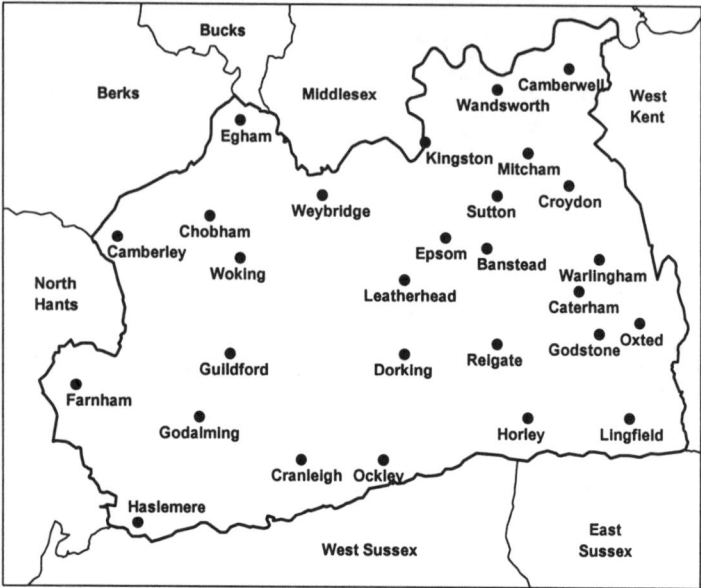

Surrey in relation to bordering vice-counties

GEOLOGY AND BIOGEOGRAPHY

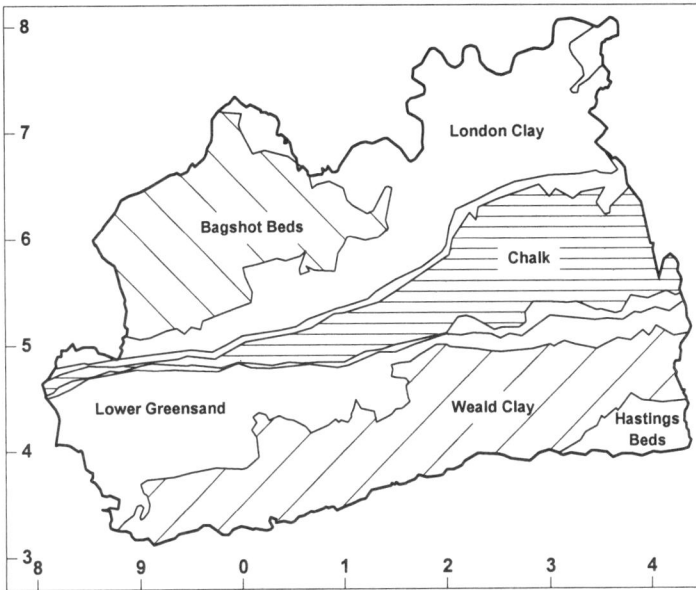

Solid geology of Surrey, after *'Butterflies of Surrey'*

Surrey, unlike many counties, still retains a high proportion of semi-natural habitat. To a great extent this reflects the geology, which comprises a mixture of poor sandy soils supporting heathland and species-poor woodlands, thin chalk soils best suited to grassland, thick wet clays where woodlands predominate, and alluvial soils along the main river valleys where most agriculture is concentrated. There are a number of accounts of the geology of Surrey, but perhaps the most useful is by Stevens in Lousley (1976), whose biogeographical account is also excellent. The hard rock geology is well described by Sutcliffe (in Collins, 1997) but, as will be seen in due course, this only has a partial bearing on species distribution as superficial deposits are the major factor over much of the county.

Soil type and related underlying geology are important factors in biogeography. Most of us will be familiar with pH which can play such an important part in determining the distribution of some plants; calcicoles such as common rock-rose, *Helianthemum nummularium*, and calcifuges such as heather, *Calluna vulgaris*, are good examples. However, drainage and stress tolerance can be equally important. This is reflected in the distribution of sand spurrey, *Spergularia rubra*, which is mainly confined to sands and gravels, and marsh thistle, *Cirsium palustre*, which predominates on clay and soils with impeded drainage.

Where plants are host to a particular invertebrate, their own distribution will clearly dictate that of the associated species. Some good examples amongst the hoverflies include *Cheilosia variabilis,* which is associated with the figworts *Scrophularia nodosa* and *S. auriculata,* and *Portevinia maculata* which is associated with ramsons, *Allium ursinum.* Not all invertebrate-plant associations are known, however, and interpretation of some distribution maps must be a matter of conjecture. Some useful pointers could be provided by referring to maps of plants, but Lousley (*loc. cit.*) does not illustrate all species and, unfortunately, misses out a substantial number of the commoner species. Thus, an opportunity to make progress on the biology of some phytophagous hoverflies may have been lost. There is an urgent need to address this issue, because mapping projects should not be seen in isolation; in due course the relationship between particular species may be easier to establish given appropriate recording.

The range of lifestyles adopted by invertebrates complicates the picture still further; not all are phytophagous, and many are predatory. Unless a predatory species is host-specific, and the host has a particular habitat association, it can be very difficult to interpret the predator's distribution. Moreover, the prey species may actually favour a well-drained or impeded soil without being associated with a particular plant or vice-versa. Thus, although the underlying geology can be significant, other factors such as drift geology can play a key role in determining invertebrate distribution, where for example drainage is more important than soil type, or where heat absorption or reflection are critical factors. This is clearly illustrated by species such as *Sphaerophoria batava* and *Paragus haemorrhous* which are essentially associated with dry habitats, yet odd outliers occur on the Weald where clays predominate but there are patches of superficial gravels.

The most obvious feature of Surrey's landscape is the chalk scarp which forms a wedge roughly east to west through the centre of the county. This high ground ranges from 267 metres at Botley Hill above Woldingham to 224 metres at Box Hill and, after dipping into the Mole Gap, rises again to 226 metres at White Downs; it drops again into the Wey valley before rising up to form the Hog's Back at 152 metres. Along the scarp and dip slopes the soils are relatively thin and well drained. Many of these support calcareous grasslands which have not been improved by modern agriculture and are now in public ownership or are owned and managed by bodies such as the National Trust. The distribution of chalk scarp grasslands is reflected in the distribution of species such as *Microdon devius.* Elsewhere, superficial deposits of clay-with-flints and sands and gravels mean that the calcareous influence to the north is more limited, yet its impact is still apparent in the distribution of such species as *Cheilosia barbata* and *C. soror.* Where the cap is clay-with-flints the soils are deeper and wetter so, for example, damp woodlands occur to the east around

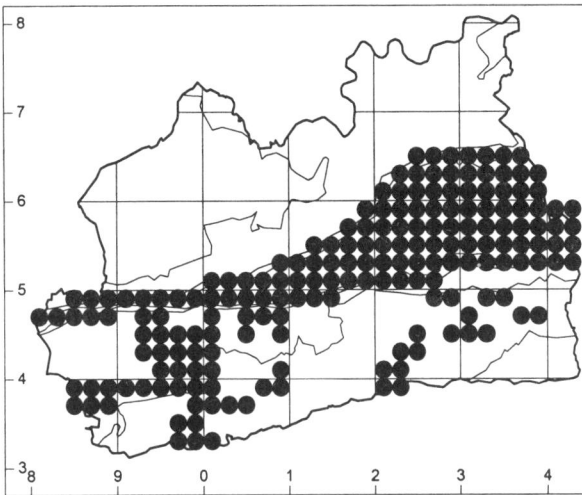

Calcareous soils in Surrey

Chelsham. Heathland also occurs sporadically, as at Headley and Banstead Heaths where the deposits are principally sands and gravels.

South of the Chalk, a band of Lower Greensand runs east from the south-west corner of Surrey where the heathlands of Thursley, Hankley and Frensham Commons cover an extensive area. The Greensand also forms the high ground of the Hurtwood, Winterfold Forest and Leith Hill, the highest point in Surrey at 292 metres. This high ground is mainly covered with stunted oak woodlands with an understorey of bilberry, *Vaccinium myrtillus,* and supports an impoverished hoverfly fauna perhaps reflecting the more limited range of plants in this area. The lower ground in this area is often heathland, heavily scrubbed up with conifers, which supports many of the typical heathland species such as *Sphaerophoria batava* and *S. philanthus.* Other heathland species such as *Microdon analis* and *Pelecocera tricincta* appear to be absent. Where the native woodlands and heathlands have been replaced by conifer plantations, a distinct fauna is apparent, exemplified by the distribution of *Parasyrphus annulatus* which is mainly confined to this discrete area.

The clays of the Low Weald are calcareous, generally poorly drained and difficult to work for agriculture, and consequently much of the Weald remains richly wooded. These woodlands are characterised by deep stream gorges along which ramsons is often profuse. The south-eastern corner of Surrey is more intensively agricultural with extensive arable and improved pasture. This is reflected in the distribution of a number of species such as *Epistrophe grossulariae*, a hoverfly whose larvae prefer arboreal aphids (G. Rotheray,

7

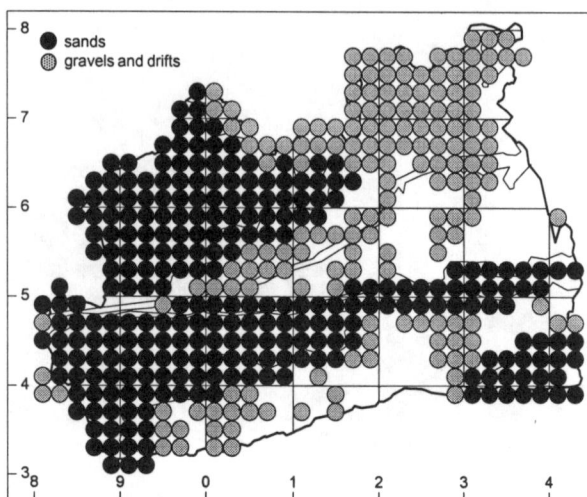

Sands, gravels and drifts in Surrey

pers. comm.). This species seems to be curiously absent from much of south-east Surrey, perhaps reflecting a preference for more heavily wooded areas. There are other species such as *Platycheirus granditarsus* and *P. rosarum* which are much commoner in this area and reflect the presence of wetter ground, damp ditches and small ponds.

The Bagshot Sands and Bracklesham Beds cover much of north-west Surrey and comprise fine sands and podsols with a clay element in places. These support a further extensive tract of dry and wet heathland, indicated by the presence of such species as *Microdon analis* and *Pelecocera tricincta* on dry heath, and *Platycheirus occultus* in wetter areas.

Further east, the London Clay predominates. Unlike the wealden clays, this is not calcareous thus in part explaining why some species which occur in the Weald are not well represented north of the Downs. The situation is confused by the presence of a series of terraced river gravels overlying the London Clay to the east; thus some heathland species such as *Sphaerophoria fatarum* and *Trichopsomyia flavitarsis* are found in parts of London, such as Wimbledon Common and Richmond Park respectively, where heathy habitats occur. These examples clearly indicate the impact that drift can have on interpretation of some of the distribution maps.

Finally, the spread of the London suburbs is the major factor behind hoverfly distribution in Surrey. Today, much of north-east Surrey is covered by urban sprawl, with small pockets of open space such as Richmond Park and Wimbledon Common. Major conurbations in west Surrey are also expanding,

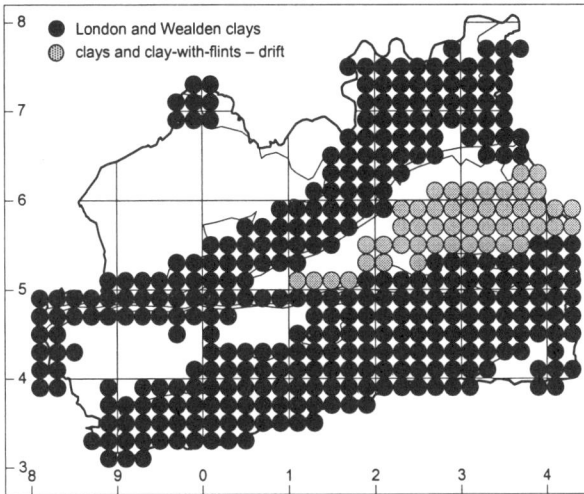

Clays in Surrey

but at the moment areas south of the North Downs are still relatively free of urban expansion. These effects are clearly demonstrated by the paucity of records from much of the London area and the absence of many otherwise common species such as *Rhingia campestris,* whose larvae feed in cow dung and which is consequently scarce.

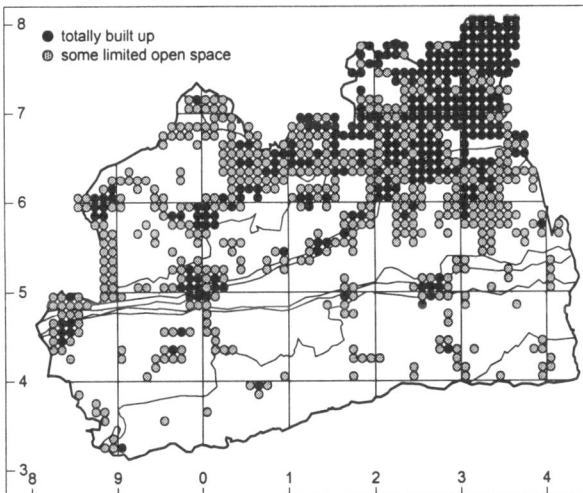

Major urban areas in Surrey

HOVERFLY BIOLOGY AND FORM

Hoverflies (Syrphidae) are very variable in form and size, ranging from the tiny *Neoascia* and *Paragus* to relative giants such as *Volucella zonaria*. Most people will be aware of those species which are promoted as the gardener's friend because they are a natural defence against aphids. Some, such as *Syrphus* spp., will be familiar to the gardener as yellow-and-black wasp mimics, some have a range of spots and others are completely black. Many are flower-visitors, hence their American name of Flower Flies, but some, such as *Xylota segnis,* rarely visit flowers at all. They are an attractive group but include a number of taxonomically complicated genera which have often made identification difficult in the past.

Whilst the adult is the most noticeable part of the hoverfly's life history, most of its time is spent in the egg, larval and pupal stages. Some, especially the predatory species, have a relatively short life-cycle with perhaps two or three generations per year; others, such as those whose larvae feed in association with rotting wood, may take a year or many years to pass through the larval stage. Larval feeding strategies are very variable, occupying most of the obvious niches; some are predators of ant larvae, of aphids (some are thought to be associated with aphids in ant nests) or of beetle or moth larvae; others feed in dung, decaying vegetation and perhaps decaying carrion; there are also a great many plant feeders which mine leaves or tunnel through roots, stems and bulbs; some feed on fungi, especially those of the genus *Boletus*, but are also thought to breed in truffles or other subterranean fungi; a further group feed in association with decaying timber, on fungi which cause rot or are associated with sap runs, or on microscopic organisms in the soup which forms in cavities in trees. The list is considerable and many species are extremely specialised, making it possible to use them as habitat indicators.

This range of form and life history makes the hoverflies an attractive group to study and one which can be used by the conservation manager to ensure that appropriate strategies are adopted to maintain and enhance the range of organisms that occur on a site. Our knowledge of the group is continually improving, but there is plenty of scope for further study (see Biological Observations, page 36).

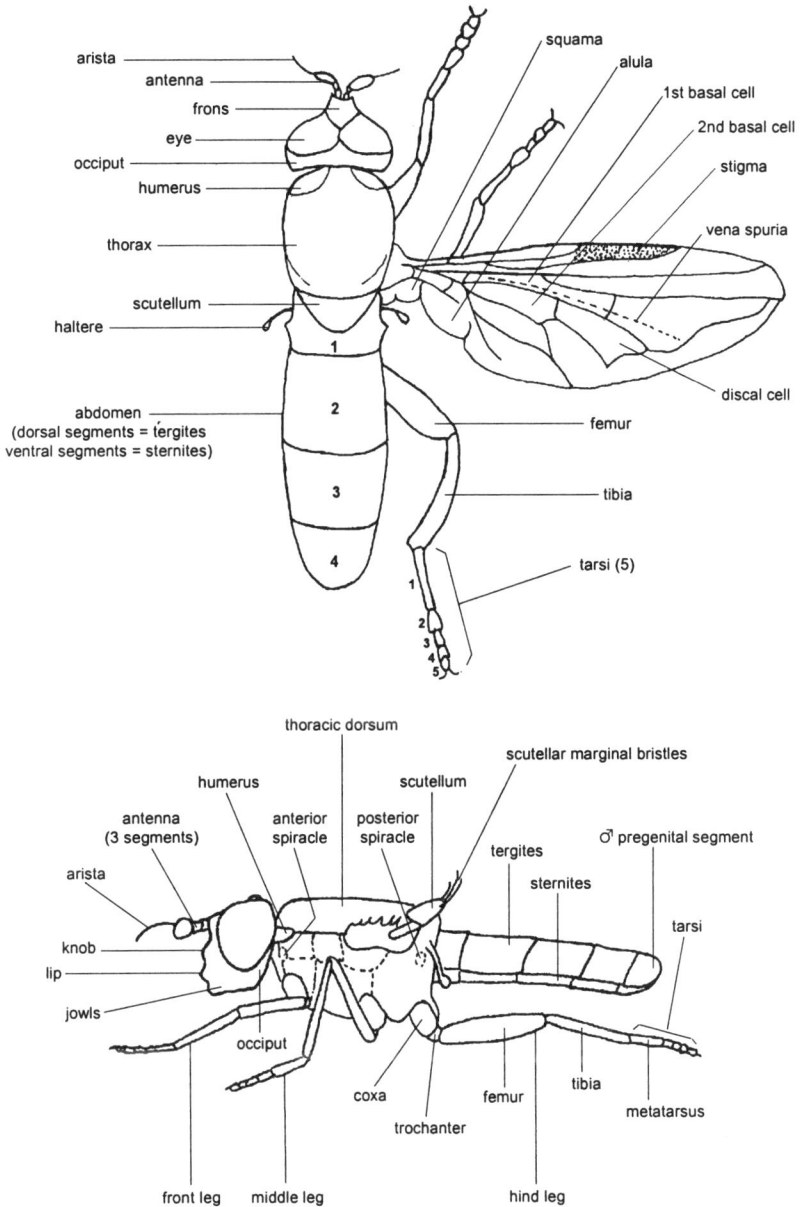

Figure 1. **Hoverfly morphology**

HABITAT ASSOCIATIONS

There are a number of obvious trends in hoverfly distribution which will be remarked upon in the main species accounts, but there would also seem to be merit in drawing attention to apparent assemblages. As part of this process, I have expanded the concept of defined indicator species started by Stubbs (1982) and developed by Whiteley (1987). Each indicator is graded 1 to 3, being the strongest and weakest respectively. The ideas developed here are, like earlier ones, intended to provoke debate and perhaps more careful recording to develop the theme. It must be stressed that they have been developed in a Surrey context and should not be used unconditionally elsewhere in the UK.

CHALK

Surface exposures of chalk range from a wide band at the eastern end of the county to a narrow ridge at the western end (the Hog's Back). As discussed earlier, much of the Chalk is capped with clays, sands and gravels, and this complicates interpretation of species' distribution where the underlying geology is chalk. There would appear to be three distinct groups of hoverflies associated with the Chalk:

Chalk grassland/woodland edge species such as *Microdon devius* and *Cheilosia nigripes*. The former is particularly sensitive to scrub encroachment and is threatened over much of its range; the latter is widespread on the Chalk and seems to be particularly associated with areas where chalk is not overlain with clay.

Woodland species which are particularly well represented along the Chalk which is generally well wooded beyond the south-facing scarp. Obvious indicators of such woodlands include *Brachyopa pilosa*, *Volucella inflata* and *Brachypalpus laphriformis*, whilst common woodland species such as *Criorhina berberina* and *Ferdinandea cuprea* are clearly well established throughout the chalk woodland complex.

Chalk associates whose distribution spreads over a wider area including the London Clay to the north (e.g. Bookham and Epsom Commons), Wealden Clay to the south (e.g. Chiddingfold Woods) and the Lower Greensand to the south (e.g. Friday Street). Species typical of this group include *Cheilosia barbata* which is quite widespread over all of these habitats, and *Cheilosia soror* which is restricted to a much narrower band of chalk and adjacent clay. *Doros profuges* seems to be closely associated with the Chalk and nearby clay at Epsom Common, and is confined to a remarkably small area. Others, such as *Xanthogramma citrofasciatum*, are commoner on the Chalk than elsewhere but are seemingly associated with dry sites similar in structure to that of chalk grassland. The following is a list of possible indicators of calcareous influences:

Platycheirus manicatus	C3
Chrysotoxum elegans*	C2
Doros profuges*	C2
Xanthogramma citrofasciatum*	C3
Cheilosia barbata	C2
C. nigripes	C1
C. soror	C1
Microdon devius*	C1
* Ant associates, possibly only loosely associated with Chalk	

Proposed chalk indicators for Surrey

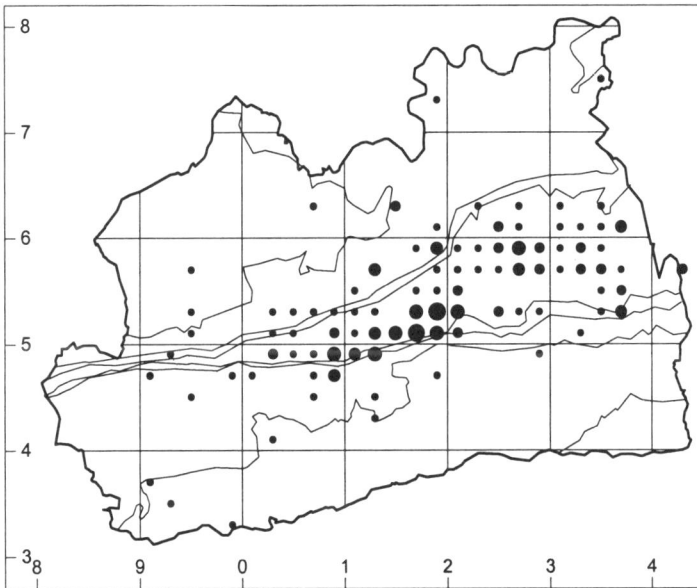

Coincidence of chalk associates (C1 & C2 only)

HEATHLANDS

The main Surrey heaths include Chobham Common, Pirbright and Ash Ranges, and Horsell, Wisley, Thursley, Hankley and Frensham Commons. They broadly comprise two elements: dry *Calluna* heathland on well-drained sand, and wet heath, typified by *Erica tetralix*, where there is impeded drainage. Their fauna is typical of heathland with a marked reduction in numbers of species, but with some characteristic associates such as *Pelecocera tricincta*, *Microdon analis, Cheilosia longula, Sphaerophoria batava, S. philanthus* and *S. virgata*. In central Surrey the heathlands of Hurtwood, Esher Common and Headley and Reigate Heaths support a more restricted but characteristic fauna such as *Parasyrphus vittiger* and *Dasysyrphus tricinctus* which are particularly frequent on heathland invaded by pine.

The following is a first attempt at a list of heathland indicators, based on the Surrey fauna:

H	**heathland indicator**	*Platycheirus occultus**	H1	-	WH2
		Paragus tibialis	H1	-	C1
C	*Calluna* **indicator**	*Chrysotoxum octomaculatum*	H1	-	C2
		*Dasysyrphus pinastri***	H3	HP1	-
WH	**Wet heath indicator**	*D. tricinctus***	H2	HP2	-
		*Didea fasciata***	H3	HP2	-
HP	**Conifer invasion/ plantation indicator**	*D. intermedia***	H2	HP1	-
		*Eupeodes nitens***	H3	HP2	-
		*Parasyrphus malinellus***	H3	HP1	-
		*P. vittiger***	H2	HP1	-
		*Scaeva selenitica***	H2	HP1	-
		Sphaerophoria batava	H2	-	C2
		S. fatarum	H1	-	C1
		S. philanthus	H1	-	C1
		S. virgata	H1	-	C1
		Cheilosia longula	H2	-	C3
		C. scutellata	H3	HP2	-
		Chrysogaster virescens	H1	-	WH3
		Melanogaster aerosa	H1	-	WH2
		Neoascia tenur	H3	-	-
		Orthonevra geniculata	H2	-	-
		Pelecocera tricincta	H1	-	C1
		*Trichopsomyia flavitarsis****	H2	-	-
		Sericomyia lappona	H2	-	WH2
		S. silentis	H3	-	WH3
		Microdon analis	H2	-	-
		M. mutabilis	H2	-	WH3

* This species is associated with poor and rich fen elsewhere in Britain (S. Falk, *pers. comm.*), but has not been found in such locations in Surrey.

** Species which are really indicative of conifers which are often a major component of heathland, albeit unwanted in many instances.

*** Associated with Psyllid galls on *Juncus*.

Proposed heathland indicators for Surrey

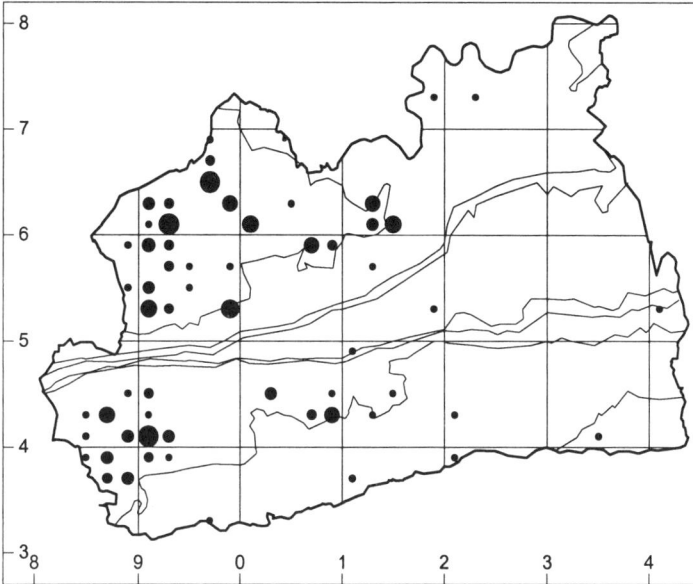

Coincidence of heathland associates (H1 & H2 only) excluding those linked to conifers

WETLANDS

Wetlands favoured by hoverflies are scarce in Surrey. Many ponds suffer heavy visitor pressure and have little emergent vegetation, while others are urbanised with concrete banks. Thus many water bodies support a limited suite of wetland species, even though dragonflies may be moderately well represented, as demonstrated by Follett (1996). There are a few major water bodies such as Bay Pond and Hedgecourt Lake which have rich fen vegetation on some banks. Here the suite of *Platycheirus*, *Anasimyia*, *Parhelophilus*, *Orthonevra* and *Riponnensia* is well represented.

In north-west Surrey the gravel pits such as those at Thorpe have margins of reed swamp and are a stronghold for *Tropidia scita*, a species which is otherwise scarce in Surrey. In many instances, however, planning conditions require that these pits are returned to agriculture, so the long-term benefit to nature conservation is limited.

Most of the remaining wetlands of interest are those associated with the canals and rivers of west Surrey, for example the wet meadows and woodlands along the Wey Navigation (Canal). In places, the canal system has fallen into disrepair and has silted up. Such places are often good sites for wetland hoverflies but, with the emphasis on restoring the canals, many of these sites have disappeared during the course of this survey, although some marginal vegetation remains in places.

Some of the best fen and carr vegetation is along the River Tillingbourne, with particularly nice examples at Gomshall (TQ0847) where *Orthonevra brevicornis* has been taken along with *Meligramma guttatum*. *O. brevicornis* also occurs at Tuesley (SU968421), a superb flushed valley fen which in addition to supporting a rich wetland hoverfly fauna is one of the few Surrey sites for the soldier flies *Oxycera rara* and *Oxycera nigricornis* (Stratiomyidae). A small flush system at Cobham (TQ102603), one near Wonham Mill (TQ223497) and the flushed meadows at Shalford (SU998473) are the other examples that come to mind. At Shalford the vegetation mainly comprises reed sweet-grass, *Glyceria maxima*, and is a locality for *Anasimyia transfuga*.

Surrey is important for the valley mires on its heathlands, but water shortages mean that these are in a highly precarious state. We can already be fairly certain that *Anasimyia lunulata* has been lost, as it has not been seen for some thirty years and its last known site at Thursley Common has changed substantially. Other species associated with wet heathland, such as *Sericomyia lappona,* are more secure. The wet heathland assemblage also includes *Orthonevra geniculata* and *Xylota florum*, the latter being associated with woodland on wet heath.

There is also a suite of more ubiquitous wetland hoverflies, such as *Platycheirus granditarsus, P. rosarum* and *Eristalinus sepulchralis,* which seem to favour the clays north and south of the Chalk. These have more catholic requirements and are often associated with wet meadows, drainage ditches and, in the case of *E. sepulchralis,* eutrophic rivers and streams, especially in the London area.

The following list of possible wetland indicators in Surrey draws on the ideas proposed by Whiteley (1987), and is modified in the light of recording in Surrey. Whilst trying to interpret the distribution of some wetland species, clear differences in the distribution of closely related species became apparent, for example *Neoascia tenur* and *N. meticulosa*, and *Parhelophilus frutetorum* and *P. versicolor*. The factors behind these differences are far from clear and it would be useful to examine in far more detail those localities where such species occur.

Code		Species	W	AW	CW	EW	RM
W	Wetland indicator	*Platycheirus fulviventris*	W1	-	-	-	RM1
		P. occultus	W1	AW1	-	-	-
AW	Acidic wetland indicator	*P. granditarsus*	W2	-	CW2	-	RM2
		P. rosarum	W2	-	CW1	-	RM2
		Cheilosia albipila	W3	-	CW3	-	-
CW	Clay wetland indicator	*Chrysogaster cemiteriorum*	W3	AW3	CW3	-	-
		C. virescens	W1	AW1	-	-	-
		Lejogaster metallina	W3	-	-	-	-
EW	Eutrophic wetland indicator	*Melanogaster aerosa*	W2	AW1	-	-	-
		M. hirtella	W3	-	-	-	-
		Neoascia geniculata	W1	-	-	-	RM2
RM	Rich marginal vegetation indicator	*N. interrupta*	W1	-	-	-	RM1
		N. meticulosa	W2	AW2	-	-	-
		N. obliqua	W1	-	-	-	-
		N. tenur	W2	AW1	-	-	-
		Orthonevra brevicornis	W1	-	-	-	RM2
		O. geniculata	W1	AW1	-	-	-
		O. nobilis	W3	-	-	-	RM3
		Riponnensia splendens	W2	-	-	-	RM2
		Anasimyia contracta	W1	-	-	-	RM2
		A. lineata	W1	-	-	-	RM2
		A. lunulata	W1	AW1	-	-	-
		A. transfuga	W1	-	CW2	-	-
		Eristalinus sepulchralis	W3	-	-	EW2	-
		Helophilus hybridus	W2	-	-	-	RM3
		Parhelophilus frutetorum	W1	AW3	-	-	RM1
		P. versicolor	W1	-	-	-	RM1
		Pipiza lugubris	W3	-	-	-	-
		Trichopsomyia flavitarsis	W2	AW1	-	-	-
		Sericomyia lappona	W3	AW2	-	-	-
		Tropidia scita	W1	-	-	-	-
		Xylota florum	W3	AW2	-	-	-

Proposed wetland indicators for Surrey

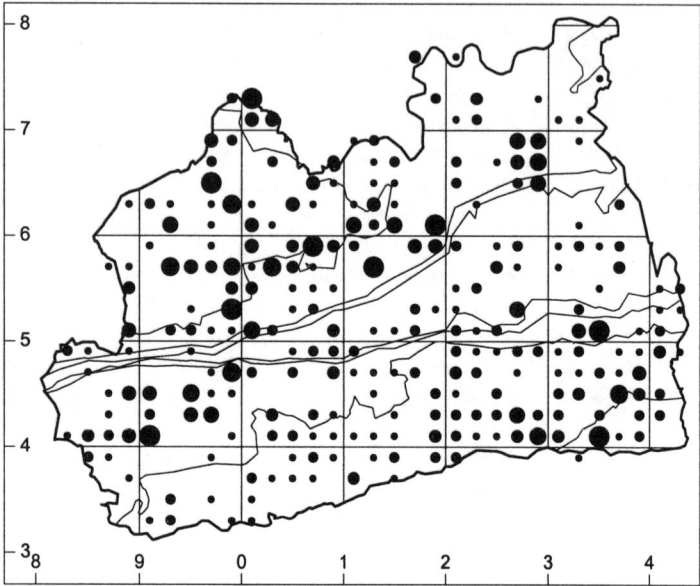

Coincidence of wetland associates (W1 & W2 only)

WOODLANDS

Woodland hoverflies can be divided into those which are clearly associated with timber, both live and decaying, those associated with woodland aphids and those which mine woodland plants. There are a few exceptions such as *Rhingia rostrata,* whose precise life-style is unknown but it may be associated with badger dung or carrion. The best woodland indicators appear to be those associated with decaying timber, and many are clearly concentrated on the Chalk and into the Low Weald. Many aphidophagous 'woodland' species are probably a great deal more mobile and catholic in their habitat preferences, including suburban gardens. Moreover, it seems unlikely that many aphid predators are entirely host-specific, making such them poor indicators of ancient woodland.

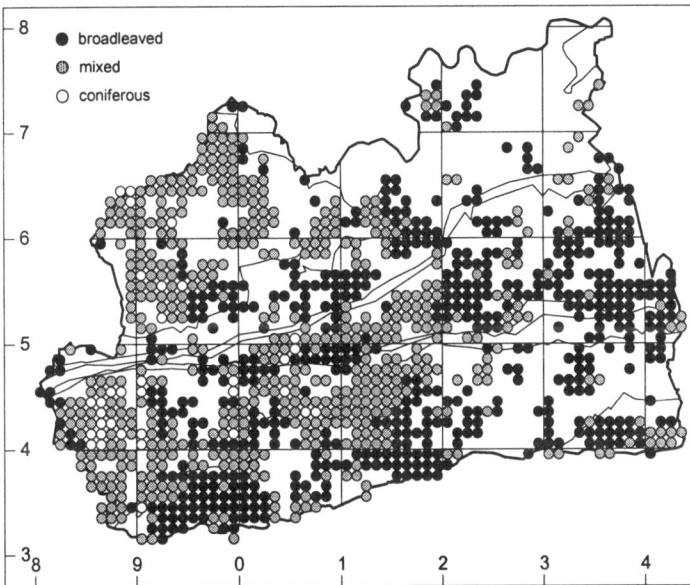

Woodland cover in Surrey – a map based on the presence of significant blocks of woodland as shown on recent OS maps

Old woodland indicators

There have been a number of proposals for woodland indicators (Stubbs, 1982; Whiteley, 1987) and with time these may be refined. Judging from the maps in this atlas and others, indicators of ancient woodland are probably impossible to define.

Surrey is noteworthy for the level of ancient woodland cover (see Drucker,

Whitbread & Barton, 1988) although it is clearly scarce in the London area (Spencer, 1986). This is actually a hindrance to identifying old woodland indicators because recording levels are still low and the distribution of most species is patchy, making it difficult to correlate with ancient woodland distribution. The following proposed indicators of high-quality semi-natural woodland comprise known saproxylic (dead wood) species and a few phytophagous species:

OW	Old woodland indicator				
		Callicera aurata	OW2	-	-
		Cheilosia antiqua	OW2	-	ClW2
		C. barbata	OW2	ChW2	-
ChW	Chalk woodland indicator	*C. carbonaria*	OW2	-	ClW2
		C. nebulosa	OW2	-	-
		C. soror	OW2	ChW2	-
ClW	Clay woodland indicator	*Ferdinandea cuprea*	OW3	-	-
		F. ruficornis	OW3	-	-
		Portevinia maculata	OW2	ChW2	ClW2
		Rhingia rostrata	OW3	-	-
		Brachyopa bicolor	OW3	-	-
		B. insensilis	OW3	-	-
		B. pilosa	OW1	ChW2	-
		B. scutellaris	OW2	-	-
		Myolepta dubia	OW2	-	-
		Sphegina clunipes	OW2	-	ClW2
		S. elegans	OW2	-	ClW2
		S. verecunda	OW2	-	ClW2
		Eumerus ornatus	OW2	ChW2	-
		Brachypalpoides lentus	OW1	-	-
		Brachpalpus laphriformis	OW1	-	-
		Caliprobola speciosa	OW2	-	-
		Chalcosyrphus nemorum	OW3	-	-
		Criorhina asilica	OW2	-	ClW3
		C. berberina	OW2	-	-
		C. floccosa	OW3	-	-
		C. ranunculi	OW2	-	-
		Pocota personata	OW3	-	-
		Volucella inflata	OW2	-	-
		Xylota abiens	OW1	-	-
		X. sylvarum	OW3	-	-
		X. xanthocnema	OW2	-	-

Proposed old woodland indicators for Surrey

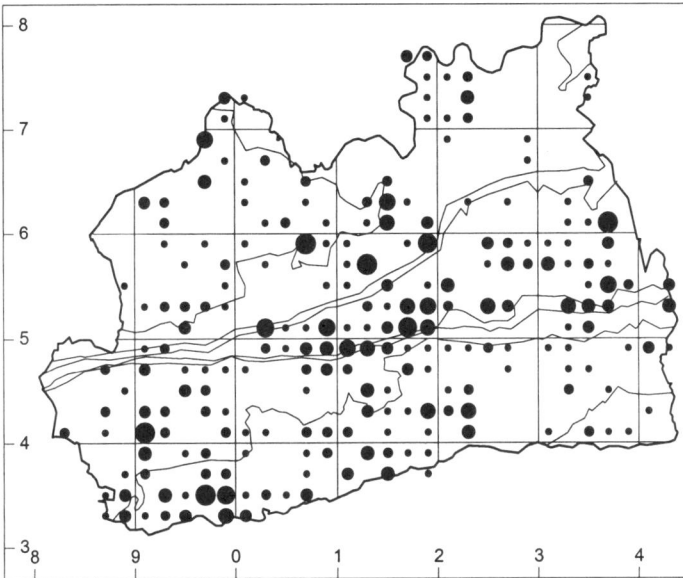

Coincidence of old woodland indicators (OW1 & OW2 only)

General woodland indicators

There are some aphid predators and a few *Cheilosia* species, associated with plants and fungi, which form the remainder of the woodland assemblage and are far more widely distributed than the old woodland indicators listed above. Some, such as *Leucozona glaucia*, *L. laternaria* and *Pipiza austrica*, are associated with hogweed aphids and are perhaps more indicative of woodland rides or edges with umbellifers. These species may be useful as indicative of woodland habitat and are listed overleaf:

W	Woodland indicator			
		Platycheirus tarsalis	W1	-	-
		Dasysyrphus pinastri	W1	-	CW1
BW	Broadleaved woodland indicator	*D. venustus*	W2	BW2	CW2
		Didea fasciata	W2	BW2	CW2
		D. intermedia	W3	-	CW1
		Epistrophe nitidicollis	W3	BW2	-
CW	Conifer woodland indicator	*Eupeodes nitens*	W2	-	CW2
		Leucozona glaucia	W2	BW2	CW3
		L. laternaria	W3	BW3	CW3
		Melangyna cincta	W2	BW2	-
		M. quadrimaculata	W1	-	CW2
		Meligramma trianguliferum	W2	BW2	-
		Parasyrphus annulatus	W1	-	CW1
		P. lineola	W1	-	CW1
		P. malinellus	W2	-	CW1
		P. vittiger	W2	-	CW1
		Scaeva selenitica	W3	-	CW1
		Cheilosia lasiopa	W2	BW2	-
		C. scutellata	W2	BW2	CW1
		C. variabilis	W2	BW2	-
		Chrysogaster solstitialis	W3	-	-
		Pipiza austriaca	W2	BW2	-

Proposed general woodland indicators for Surrey

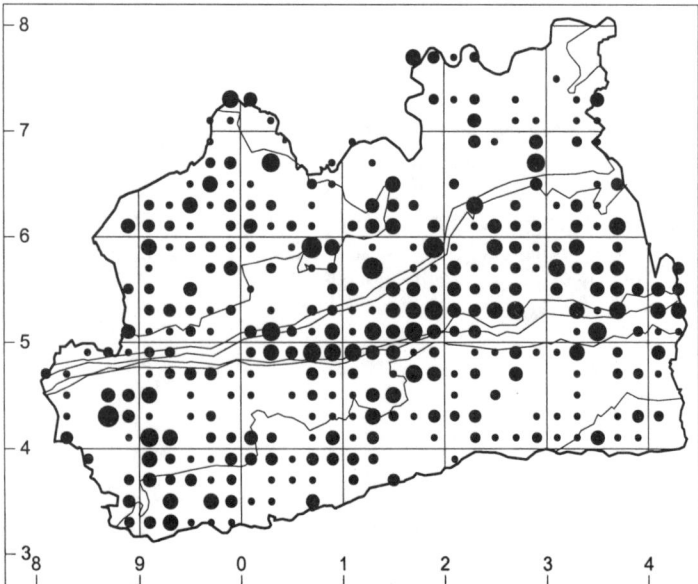

Coincidence of general woodland indicators (W1 & W2 only)

SOME IMPORTANT SITES IN SURREY

This brief account is intended to highlight some of the more interesting sites but is not exhaustive and should be used sparingly by the newcomer to hoverfly recording. Moreover, it should not be used to justify the loss of other interesting sites simply because they are not listed here; such sites are mostly represented in full by the site records linked to individual species accounts. One point I would stress is that I have never ceased to marvel that one can visit a seemingly unremarkable area and still be surprised by some of the species one encounters. In fact, I would regard it as essential for the student of hoverflies to visit such sites as car parks with adjacent ruderals, rough building sites, roadside verges and small spinneys, before judging the frequency and requirements of many species.

Ashtead/Epsom Common TQ15 & TQ16

This complex of oak woodland and neutral grassland on clay has yielded a great many important records, including a good representation of saproxylic species such as *Criorhina asilica* and *C. ranunculi*. Many of the species whose larvae develop in plant stems and roots are well represented and these include scarcities such as *Cheilosia barbata* and *C. carbonaria*, but perhaps the most remarkable record is that of *Doros profuges* in 1993, well away from its traditional downland haunts. Much of the woodland is comparatively recent, but there are small numbers of extremely old oak pollards in more remote areas. At the Epsom end, the large artificial ponds provide interesting variation and habitat for species such as *Neoascia interrupta* which are associated with aquatic and semi-aquatic habitats.

Bay Pond TQ35

This Surrey Wildlife Trust (SWT) reserve includes a substantial area of rich fen and alder carr on its northern banks. It is one of the best wetland localities in Surrey with a sizable hoverfly list and a good assemblage of wetland species such as *Anasimyia* spp., *Parhelophilus* spp. and *Neoascia* spp.

Box Hill TQ15

This is a remarkable mosaic of chalk downland on the scarp and dip slopes, which supports dry grassland species such as *Xanthogramma citrofasciatum* and *Microdon devius*, and beech hangers on deeper soils with a cap of clays and gravels, which have yielded woodland indicators such as the four *Criorhina* species, *Melangyna quadrimaculata* and *Brachypalpus laphriformis*. Box Hill is part of a complex of woodland and grassland in central Surrey which has yielded a higher number of species than anywhere else in the county and is clearly a hot spot even in UK terms. This site complex includes Mickleham Downs and Headley Warren and has, historically, been the main locality for *Doros profuges*. It is an area where there would be merit in examining rot-hole faunas for species such as *Pocota personata* and *Mallota cimbiciformis*.

Chalk scarp – Oxted to Dawcombe and Ranmore to Hackhurst

The chalk scarp retains much of the best remaining habitat in Surrey. It forms a narrow band of small chalk grasslands interspersed with scrub and woodland. Much has become heavily scrubbed over during the last 40 years and, despite increasing efforts to clear scrub over the last few years, a great deal remains to be done. Around 325 hectares of chalk grassland remain in Surrey, making these slopes especially important because they form an almost contiguous run of semi-natural habitat across much of the County. Many of the species represented at Box Hill also occur on these slopes.

Chiddingfold Woods SU93

Many of the Chiddingfold woods are now coniferised, but the remaining coppice-with-standards woodlands are still of considerable interest for their saproxylic fauna. This is clearly illustrated by the distribution of the typical assemblage of *Criorhina* spp, *Brachypalpoides lentus*, *Xylota abiens*, *X. sylvarum* and *Ferdinandea cuprea*. These woods on wealden clays are characterised by the deep-cut streams whose banks are frequently clad in ramsons which supports *Portevinia maculata*. Some of the more remarkable records include those for *Cheilosia barbata* which is mainly found on the Chalk and adjacent clays, *Cheilosia carbonaria* which is mainly found in old woodland localities, and a single record for *Cheilosia soror* which is well away from its usual haunts on the Chalk. Where coniferisation has taken place there are further points of interest such as the discovery of *Microdon analis,* which would seem to be associated with the wind-throw from the 1987 storm. There is considerable scope for further recording in the Chiddingfold area and there is a high likelihood of interesting discoveries; species to look out for might include *Xylota coeruleiventris* and *Eriozona erratica*.

Chobham Common SU96

The *Calluna* and *Erica* heaths of west Surrey are particularly difficult to work because they have a tendency to be very hot and dry; this feature is not favoured by many flies, although there are some which specialise in such conditions. There is a distinct assemblage of heathland species which is well represented on Chobham Common. Perhaps the most interesting records are those for *Pelecocera tricincta* which are amongst the most northerly records for the species in Great Britain. Typical wet heathland hoverflies include *Chrysogaster virescens, Melanogaster aerosa* and the newly separated *Platycheirus occultus,* which are associated with wetter areas of *Erica* heath. The presence of pine scrub contributes such species as *Parasyrphus vittiger,* whilst characteristic heathland *Boletus* fungi support *Cheilosia longula* which is apparently more abundant on heathland than elsewhere.

Great Bookham Common TQ15

This is one of the classic Surrey localities which has been immortalised by the work of Len Parmenter from the 1940s onwards (Parmenter, 1950a; 1960; 1966). It comprises oak woodland on clay with open damp meadows and a series of ponds invaded by bulrush, *Typha latifolia*. Although this remains one of the best sites for hoverflies, work on the ponds in the 1960s means that the wetland element is not as rich as it was (A. Stubbs, *pers. comm.*). The overall species list of 133 ranks as high as the majority of other well-recorded sites across southern England and is probably higher than most, except for Windsor Great Park and the New Forest, both of which are much bigger and support greater habitat diversity. Despite the intensity of recording here, further additions are possible; for example there are no records of *Sphegina* on a site with substantial woodland cover. There are also records of two species, *Cheilosia vicina* and *Platycheirus scambus*, which I have rejected because their national distribution is northern and western and they are extremely unlikely to occur in south-east England. The full species list for this site is provided in Appendix 4.

Hedgecourt Lake TQ3540

This is another SWT reserve which supports a rich mixture of alder carr and fen at its western end. The wetland fauna is well represented, but there are also records of a number of unusual species such as *Xylota florum* which is likely to be associated with wet timber, and *Orthonevra geniculata* which is otherwise known only from the wet heaths in west Surrey.

Merrow Common TQ0251

This woodland is included as an indication of how important small isolated sites can be for invertebrates. Merrow Common is remarkable because it has yielded a higher number of nationally scarce species in relation to the level of recording than any other site in Surrey. These include *Myolepta dubia*, *Meligramma euchromum*, *Brachyopa pilosa* and *Epistrophe melanostoma*.

Richmond Park TQ17 & TQ27

One of the top sites in Great Britain for saproxylic beetles, Richmond Park has not been as diligently recorded for flies and offers considerable prospects for remarkable discoveries. It is a well-known site for *Psilota anthracina,* and both Surrey records for *Pocota personata* are from adjacent sites. Among the more unusual records from Richmond Park is that of *Trichopsomyia flavitarsis* which is usually associated with upland bogs and is a good indicator of acidic habitat. Further survey here is highly desirable.

Thursley Common National Nature Reserve & Hankley and Frensham Commons SU94 & SU84

This is one of the classic heathland complexes for which Surrey is renowned and one which is especially important for dragonflies (Follett, 1996). It is a mixture of wet and dry heath, acid pools with *Sphagnum,* large open water bodies with reed fringe, and coniferous woodland. Hankley Common is a military training area, but most of the rest is open to the public, being owned by English Nature and the National Trust. This heathland complex is the only one from which *Anasimyia lunulata* has been recorded in Surrey, and is also the only known area in Surrey for *Chrysotoxum octomaculatum,* one of the two hoverflies listed on the short list for the UK Biodiversity Action Plan (D.o.E., 1995). The mixture of wet and dry heath has yielded such heathland specialities as *Melanogaster aerosa* and *Microdon analis,* which is associated with the black ant *Lasius niger* in pine stumps and is widespread on the Surrey heaths. It is also one of the best areas in Surrey to see the spectacular robber fly *Asilus crabroniformis* (Asilidae).

Tilford Reeds SU84

This former heathland is now heavily coniferised. There was evidence of the heathland nature of this site at least until the late 1980s when I last visited it. This is a remarkable locality because of the presence of such heathland associates as *Sphaerophoria virgata* together with pine-dwellers such as *Didea intermedia* and saproxylics such as *Criorhina asilica.* Spring-flowering sallows, *Salix* spp., have yielded some of the few modern records of *Platycheirus discimanus* and *Melangyna barbifrons.*

Tuesley SU9642

This is a superb flushed valley fen with extensive areas of great horsetail, *Equisetum telmateia,* which has yielded hoverflies such as *Orthonevra brevicornis* and *Xylota florum.* It is also one of the best sites in Surrey for soldier flies (Stratiomydae), having yielded *Oxycera rara* and *O. nigricornis.*

Wisley Common TQ05

This wet heathland has been severely damaged by the M25 and widening of the A3(T) in recent years. Bolder Mere, the lake to the east of the A3, was formerly interesting for its fringe of mire vegetation, but windsurfing has led to serious trampling and there are signs of eutrophication with the arrival of bulrush, *Typha latifolia*. This site was worked for many years by Alan Stubbs and was renowned for its cranefly fauna amongst others. The hoverfly fauna includes both *Microdon analis* and *M. mutabilis*, and *Didea intermedia* which is likely to be associated with the pine that has invaded much Surrey heath. There can be little doubt that this site is on the decline and its cranefly fauna has suffered considerably, but it should still yield many of the specialist heathland hoverflies.

RECORDING HOVERFLIES IN SURREY

There has been a history of recording hoverflies in Surrey which extends well back into the late 19th Century, with frequent references to important Surrey locations by Verrall (1901). We owe a great deal to the late Len Parmenter who visited many Surrey sites and in particular recorded the faunas of Great Bookham and Limpsfield Commons as part of the work undertaken by the London Natural History Society (Payne, 1970); his collection now resides in the Natural History Museum of London. The fauna of the Surrey heathland commons has received considerable attention over the years, not least from O.W. Richards who, in addition to recording aculeate Hymenoptera, has left a legacy of records of many of the typical heathland hoverflies and the faunas of the once great Esher and Oxshott Commons. The work of R.L. Coe in north-east Surrey should not be overlooked; his records, such as those for Selsdon Woods, extend into the 1960s and provide a valuable historical perspective. Other useful historical sources include the Diver collections at Exeter Museum and at the Institute of Terrestrial Ecology at Furzebrook.

Peter Chandler and Alan Stubbs accompanied Cyril Hammond on many forays in the 1960s and 1970s and have recorded widely throughout the county (Stubbs, 1981). Alan and Peter have provided a valuable historical record of Wisley and Thursley Commons amongst others. Since then Andrew Halstead has recorded regularly at sites such as The Sheepleas, in addition to Wisley Gardens (Royal Horticultural Society), and has contributed a stream of interesting records from other sites. Roger Hawkins has been another major contributor in recent years and has provided many valuable records, especially those of flowers visited by hoverflies. Sadly, the county was robbed of one of its most active naturalists with the premature death of Rupert Hastings who was a keen hoverfly recorder with a special interest in the fauna of Kew Gardens; his records crop up regularly in this account.

Surrey was formerly a regular venue for field trips by the South London Entomological and Natural History Society (SLENHS). Their Proceedings for the 1950s and 1960s include regular reports for visits to such places as Cosford Mill, Thursley Common, Wisley and Ockham Commons, Hackhurst Downs, The Sheepleas and Oxshott Common. Some records on the Parmenter card index included in this account need to be treated with some caution as it was common for the party to assemble at a railway station and then proceed to a nearby site. The records may, however, refer to the station and not the site. The change from the SLENHS to the British Entomological and Natural History Society (BENHS) in 1968 was followed by a move towards visits much further afield, and regular trips to Surrey seem to have halted around 1973. Furthermore, the advent of the mercury vapour moth trap and wider ownership of the motor car also meant that fewer entomologists participated in field visits. Consequently, the journals become less useful as a source of records.

Changing patterns in attendance at indoor meetings of the BENHS, with an overall decline in numbers and exhibits, also mean that a further rich source of records diminished to a level where there is little profit in detailed searches for records. This is to be lamented as recorders become more insular, but a positive improvement has been seen with the advent of computerised recording schemes and a corresponding interest in submitting records. Many records are available to the County Recorder that would hitherto have been inaccessible, but the opportunity to seek useful comments on behaviour and distribution has perhaps been lost for ever.

Work on the county mapping scheme began in 1985 and in the following years Graham Collins and I have covered considerable areas of the county, often visiting five or more sites in a day. Originally, it was intended that the maps would be produced on a basis of 5km grid squares, but as trends became apparent it was clear that more refined recording was needed; this meant that the project has taken rather longer than anticipated, as good definition at a tetrad level is far harder to achieve. Much additional recording has been undertaken to ensure that coverage is as even as possible, but even so, it is clear that there are gaps which are unavoidable. These provide future recorders with an indication of the effort required to get an adequate level of coverage; hopefully someone will be tempted to repeat the process in due course. Recording effort has not been even on a yearly basis. Some years have been particularly productive because the weather has been suitable, whilst in others my enthusiasm for square-bashing has waned. The bulk of the records in this atlas are from the period 1985 to 1996 and, although some literature search and chasing old records has taken place, a full account of historical records has not been prepared; the maps are intended as a snapshot of the fauna in the late 20th century. Overall recording effort from 1985 to 1997 is shown below.

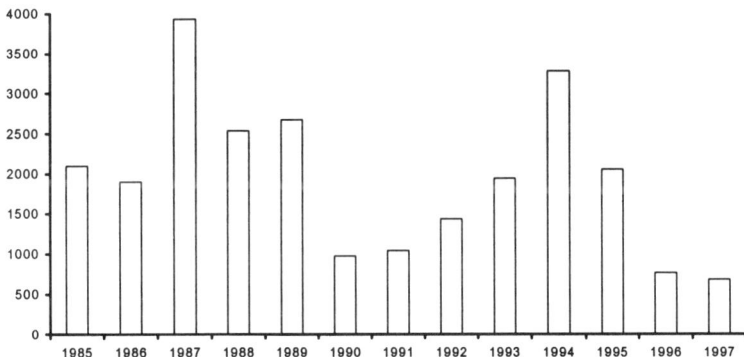

Figure 2. **Numbers of records per year, 1985 to 1997**

Most recording schemes are plagued by the tendency to record the distribution of recorders and their favourite haunts. I have therefore attempted to minimise this problem by visiting as many tetrads as possible, even if it has meant recording from roadside verges. This strategy overcomes the problem of recorder bias to some extent, but there are places where, for example, car parking is difficult or little habitat is present. This is particularly apparent within inner London, around Lingfield/Dormansland and north of Horley, in south-west Surrey around Farnham, in the garrison towns of Aldershot and Camberley, and on Pirbright and Ash ranges. Even so, 512 of the 540 tetrads comprising the vice-county have been visited on at least one occasion, and many species exhibit clear trends in distribution.

Hoverflies are mobile and sun-loving, which makes it is important to record in appropriate conditions. This factor, together with the short emergence periods of many species, means that the levels of recording achieved for plant atlases cannot be matched. Furthermore, there are few active recorders compared to ornithologists, botanists and butterfly enthusiasts. The level of coverage achieved should therefore be viewed in this context.

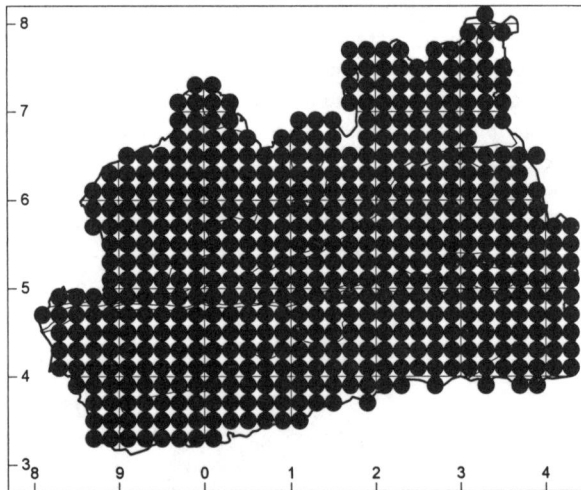

Overall coverage, all records

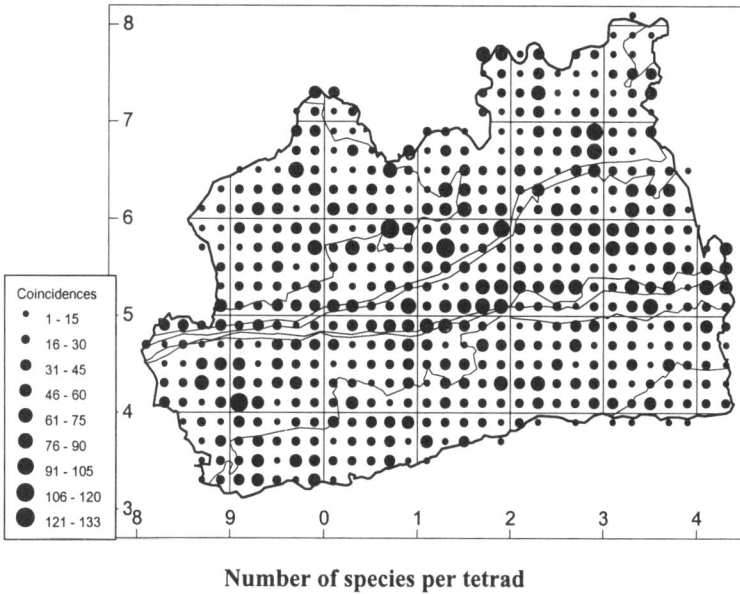

Number of species per tetrad

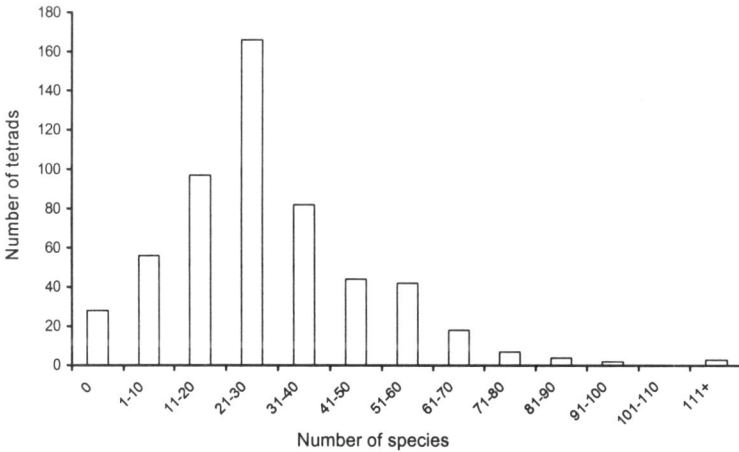

Figure 3. **Overall coverage – numbers of species per tetrad**

DATA COLLECTION, STORAGE AND RETRIEVAL

Nearly 30,000 records have been assembled for this project. They are stored on the Recorder database written by Dr Stuart Ball, which is marketed by the UK Joint Nature Conservation Committee (JNCC). This is the system used by Surrey Wildlife Trust and a copy of the data will be lodged with the Trust. The records have also been incorporated into the national Hoverfly Recording Scheme, run by Stuart Ball and myself. Maps have been prepared using the UK D-Map program (see Morton & Collins, 1992).

At this juncture, it is perhaps worth dwelling on the importance of submitting accurate and detailed data in a form that can be handled by modern computerised databases. Computerisation can be a slow process and can be greatly speeded up by submitting lists of species for a particular site on a particular date at a given grid reference point. A four-figure grid reference is essential, and even greater accuracy is a bonus. A two-figure grid reference is of limited value except to create a dot on a 10km grid, whereas a detailed record can be used, for example, to defend sites at a Public Inquiry.

Records sent as a list of species for a site over a long date range are also of very limited value. Firstly, nobody can tell whether the record was from the beginning or end of the date range, and must therefore use the earliest date; secondly, such data are of very limited value in the management and safeguarding of sites, because of their vagueness; finally, the more accurate the record, the more that can be done with the data in terms of biological interpretation and interrogation, for example identifying emergence times through phenology histograms (Ball & Morris, 1992). It should also be stressed that records arising from identification using pictorial guides are too inaccurate to be of any real use. Where necessary, and in the absence of supporting specimens, I have rejected such records both here and for the national scheme.

The tradition of publishing short notes and observations in the entomological press has resulted in a rich source of data and observations. The references used in this book clearly illustrate the value of such contributions on a regular basis by those such as Len Parmenter. I have also realised that a series of short notes on my own collecting trips might have been useful memory joggers, for it is inevitable that some information becomes clouded by time and memory loss. A hint to future recorders working on an atlas such as this would be to retain a diary of ideas and special observations in addition to basic species lists for site visits; this is something I did not do, but on reflection should have done. I would also draw attention to the need to encourage publication of records of species new to Britain for some time after their initial discovery. This is exemplified by the data available for *Volucella zonaria* which was widely reported until the mid-1950s, and these records make an important contribution to our understanding of how this species became established.

Sadly, one unfortunate comment by Fraser (1955), criticising further publication of notes on this species, may have deterred others from publishing their records. Given that some species may become established only temporarily, it would be useful to have a full record of their residency.

Some recorders may not wish to specialise in hoverflies but would like to contribute to the recording process. Collections of specimens, either pinned or unpinned, from a site with appropriate data (date, grid ref., site name) will usually be gladly received by a County Recorder who should then provide feedback on what you caught. Expect a delay however, as entomologists are notoriously slow to respond. Those hoverfly enthusiasts who are unsure about identifications of problem genera such as *Cheilosia*, *Sphaerophoria* and *Pipiza* should retain material and submit it for identification by the County Recorder.

RECORDING TECHNIQUES

There are a great many factors which will influence the range of species recorded on a field trip, the most significant being the field-craft of the recorder. Most recorders will concentrate on those species which visit common flowering plants, but there are other ways of finding hoverflies.

Many spring hoverflies bask on sunlit leaves and can be found in such locations; this includes common species such as *Syrphus*, as well as less common species. Different tree leaves have differing values to leaf baskers; sycamore, *Acer pseudoplatanus*, and hybrid lime, *Tilia* x *europea*, are particularly favoured, but horse chestnut, *Aesculus hippocastanum*, can also be productive where neither of the others are present. The following species may be sought on sunlit leaves in April and May:

Epistrophe eligans	*Syrphus* spp.
E. nitidicollis	*Cheilosia nigripes*
Melangyna lasiophthalma	*C. variabilis*
M. cincta	*Brachyopa scutellaris*
Meligramma euchromum	*Heringia heringi*
M. trianguliferum	*H. (Neocnemodon)* spp.
Meliscaeva auricollis	*Pipiza noctiluca*
M. cinctella	*P. luteitarsis*
Parasyrphus punctulatus	*Criorhina floccosa*

Searching tree trunks, especially those with sap runs, will make finding *Brachyopa* species more straightforward. Watch the sap run and await the arrival of the adults. Do not stir up the sap run as this could cause mortality to fragile eggs and young larvae. When a productive tree is found, it is also well worth holding on to a few specimens because individual trees may support a number of species which are difficult to identify in the field.

Many species of hoverfly have remarkably short emergence periods and this becomes obvious when a number of sites are visited on the same day. It is particularly worthwhile to search for scarcer species at a number of sites once they are known to be flying; *Myolepta dubia* is a particularly good example of a species with a short emergence period, but there are others such as *Brachyopa* spp., *Criorhina asilica* and *Platycheirus tarsalis*.

There are advantages to sweeping grassland for *Melanostoma* and *Platycheirus*, and woodland vegetation for *Sphegina*, and it is worthwhile retaining a good number of specimens for critical identification as more than one species may be present. This is particularly pertinent to the recently separated *P. clypeatus* group.

There are a number of species such as *Portevinia maculata* and *Neoascia obliqua* whose association with particular plants can be used to identify possible localities. Ramsons, the host plant of *P. maculata*, can be detected by smell and once found it is usually possible to find the fly in attendance. *Neoascia obliqua* is believed to be associated with butterbur, *Petasites hybridus*, and where good stands of this plant occur there seems to be a good chance of finding the hoverfly; these suspicions only came to light at the end of this survey, so detailed search of butterbur beds may reveal that it is a great deal commoner than currently believed.

Recording larvae is another possibility. Records of associations with particular plants or their associated aphids would be highly beneficial, especially as there are few published records of predator/prey associations. Collecting material from sap runs, rot-holes or decaying timber may help to establish the presence of species which are otherwise poorly known, and may also extend our knowledge of species' preferences in terms of the types of rot-holes and the trees they favour. Much useful guidance to this can be gained from Rotheray (1993).

For those who do not want to record more widely, there is also a lot that can be learnt from studying one's own garden. Records of hoverflies from urban areas are particularly difficult to obtain, and a network of recorders from such areas would be especially valuable. More detailed studies by experienced recorders could include daily counts as described by Stubbs (1991). These might help to explain some of the apparent changes in frequency which more generalised recording suggests (see account for *Rhingia campestris*, page 129).

BIOLOGICAL OBSERVATIONS

Many hoverflies visit flowers as a source of both nectar and pollen, but our knowledge of which species visit particular flowers is rather patchy and few recorders make an effort to note associated flower visits, let alone to record whether the fly was utilising pollen or nectar. Extensive lists of flower visit records are published for Belgium and include some British records (de Buck, 1990), but there do not appear to be similar lists arising from British work. During the early stages of this project, I too failed to record such visits systematically, but in latter years have attempted to do so more assiduously. Various literature records of flower visits have also been incorporated into this account in order that as broad a picture as possible is presented for flower visit records in Surrey. Much of this derives from the work of Len Parmenter whose papers are consistently annotated with flower visit records.

The lists of flower visit records are far from complete as is clear from the range of records published for Belgium. None the less, I now hold over 2,500 records of flower visits by hoverflies in Surrey, and these include records for a number of plants which are not listed by de Buck (*loc. cit.*). As might be expected, many of the records are from hogweed but, overall, there are records from 196 plants visited by 152 species of hoverfly. These include records for aggregates of records from "buttercup" and "ragwort" where precise identifications have not been made. There is considerable scope for more work on flower visits, and perhaps the lists published here will stimulate others to record from particular plants and perhaps to analyse the data in far greater detail than is possible here.

Recording larval associations is a further area where much progress could be made. Rotheray (1993) provides a good foundation for such studies, but there is a great deal more to be learnt. Breeding larvae out may also be a good way of recording some of the scarcer species about which we know so little. This is particularly true for many saproxylic species such *Mallota cimbiciformis* and *Pocota personata* which are almost undoubtedly under-recorded.

In compiling this account it has also become clear that we know very little about which species of hoverfly suffer from fungal attack. I have a few records, but this is an area where more assiduous recording would be profitable. There are also few published accounts listing hoverflies as prey items of predatory flies, and there is considerable scope to expand upon this theme. Detailed observations on the behaviour of individual species are needed and could be a useful line of research. Finally, one area which everyone could record is the quantity of eggs laid by captive females; Parmenter (1951) drew attention to the value of such records, yet as far as I am aware only Groves (1956) has responded in print and thus there must be scope for everyone to do some useful recording.

There are many projects described by Gilbert (1986) which the interested amateur can follow up. They add a new dimension to the study of hoverflies and will ultimately assist in unravelling the complicated biology of this fascinating group.

COMPARISON WITH OTHER RECORDING SCHEMES

Adjacent counties have yet to be surveyed in detail for their hoverflies, but lists do exist for the London Natural History Society recording area (Plant, 1986) and for Kent (Chandler, 1969). There are, however, a number of published atlases for other counties, against which this project can be compared. Rotheray (1979) set the pattern with an atlas of the hoverflies of Staffordshire based on 10km grid squares which, at the time, was a major advance in hoverfly mapping. Since then Whiteley (1987) provided considerable detail of the fauna of the Sheffield area, covering some 15 10km squares around Sheffield and setting the trend by providing descriptions of some of the more unusual sites; moreover, he attempted to provide lists of indicator species for woodland and wetland. This account did not, however, provide maps for more than a selected range of species on a 1km square basis. More recently, the work of Levy *et. al.* (1992), based on 5km grid squares, has demonstrated that it is possible to get financial support for such projects from local records centres and paves the way for more detailed published records for County faunas. There are a number of other projects in the pipeline and it can be expected that a range of new publications will arise in the near future.

Judging by the way Surrey stands out on the national scheme maps (Ball & Morris, *in press*), it is one of the best recorded counties for hoverflies. This project demonstrates the value of intensive recording to provide a fuller picture of which species are really rare and those which are simply hard to find. Detailed survey of other south-eastern counties would undoubtedly demonstrate more widespread distribution of species hitherto considered rare.

HOVERFLY CONSERVATION

More detailed information on invertebrate conservation may be found in Kirby (1992) whose work describes a wider range of practical management options, whilst Fry and Lonsdale (1991) discuss a variety of the issues of invertebrate conservation. There are, however, a number of general rules which if followed should make it possible to avoid damage to hoverfly populations and perhaps enhance them.

- Sites should be managed for their strengths and continuity of existing habitat.

- In woodland sites the over-riding principle has been that native tree species characteristic of the geology should be preserved, and invasive aliens removed. Ancient trees with evidence of decay, such as rot-holes, should be retained, including any foreign species such as horse chestnut, sycamore and sweet chestnut, *Castanea sativa*, as these can be valuable habitat for saproxylic species. Cut timber should be left in damp situations where it will not dry out.

- Dead-wood habitat does not only occur above ground; rotting roots are a major resource and therefore grinding out tree stumps from woodland and parkland should be resisted, since they may support one of the most important elements of the hoverfly fauna, and hundreds of larvae may be present.

- Rot-holes are an important conservation resource and should not be subjected to arboricultural practices such as filling with concrete. On sites where ancient trees occur, there would be merit in conducting a census of existing rot-holes. Useful projects might include mapping rot-hole distribution in ancient trees in parkland and other woodland sites, both across the site and within individual trees. Provision of artificial rot-holes is a further possibility; these can be made out of old 2-litre plastic bottles containing a mixture of rainwater and sawdust, and can be strapped to suitable trees.

- Trees damaged by goat moth, *Cossus cossus*, are now rare. In addition to safeguarding such trees for goat moth, the sap-runs which arise from goat moth damage are important for a suite of uncommon hoverflies, some of which seem to be more closely associated with goat moth than others.

- Sap-runs are very important and generally not indicative of disease. Trees with such sap-runs should be retained unless there is very good evidence that they are a threat to public safety. Pollarding can be a good way to prolong the life of trees, and might be used to retain those which have good dead-wood habitat or sap-runs and which might otherwise be felled.

- Woodland rides and glades are most valuable, if sufficiently wide to support a varied flora with nectar sources such as the flowers of umbels, hawthorn, sallow and alder buckthorn in sunny sheltered locations, and sunlit leaves upon which hoverflies can bask.

- Woodland edge is an important habitat and should be managed to maintain a gradual transition to grassland where possible.

- Grasslands should be managed for both their flora and their structure. Do not overlook the importance of topographic features and remember that there are a number of associations with ants which respond poorly to mowing; such species include *Microdon devius* which it is suspected may have been lost from Farthing Downs as a result of mowing.

- Heathland rides and paths with small yellow composite and tormentil flowers are important nectar sources even if the hoverfly larvae inhabit another niche such as boggy areas. These rides and paths should be maintained with a gradual transition from bare ground through short turf to taller grassy herb-rich vegetation.

- Shallow puddles and wet hollows should be retained at all times as they will support many common species even if no rare species are present.

- Rank ruderal locations have their own intrinsic value and should not necessarily be the target of modification or elimination. Ruderal umbels can be a very important nectar source.

- Short worn turf can be important as a sunning spot for some hoverflies, apart from being an important nesting site for bees and wasps.

- Although there is a growing interest in hoverflies and inevitably more material is retained for identification, there are no universal grounds for concern about collecting pressure because most scarce species have habits which make them difficult to find. It is therefore unlikely that the retention of material will seriously deplete populations. A small number of specialist species such as *Hammerschmidtia ferruginea* (a Scottish species) may be at risk because they are highly specialised and their habitats are rare and easily damaged. It should also be remembered that accurate recording does require retention of voucher material for records to be accepted, and I have rejected a number of records where they were not supported by vouchers.

HOVERFLIES AND BIODIVERSITY

In 1994 the UK Biodiversity Action Plan (DoE, 1994) was published following the Government's commitment to the Convention on Biological Diversity at Rio de Janeiro in 1992. This was followed up by the publication of the report of the Biodiversity Steering Group (DoE, 1995) which produced three lists of species for which costed Action Plans were proposed. The first tranche of Action Plans, the short list, included two hoverflies, one of which, *Chrysotoxum octomaculatum*, occurs in Surrey. A further four species recorded in Surrey, *Doros profuges*, *Eumerus ornatus*, *Pocota personata* and *Microdon devius*, are listed in the last tranche, the long list. Three out of the four are well established in Surrey and one, *M. devius*, has its UK stronghold in the county. One, *P. personata*, is only known from old records but must surely still occur in Surrey.

Microdon devius is an ant associate which occurs on chalk downland, and it is thought that *C. octomaculatum* is in some way associated with ants on heathland; further autecological studies on both are desirable and improved site management to reverse the trend of scrub invasion is essential. *Eumerus ornatus* is one of a group of root or bulb feeders and may be closely associated with a particular woodland plant. It is one species whose biology might be better known if more complete plant mapping was available. Very little is known about *D. profuges* which is probably associated with aphids in ant nests (G. Rotheray, *pers. comm.*) and would certainly merit detailed autecological studies. Finding strong populations of each would be a useful starting point. Finally, *P. personata* represents a group of generally threatened species, those associated with dead and decaying timber. Studies of the rot-hole fauna of Surrey would be highly worthwhile and might help to throw light on the true status of a range of species such as *Mallota cimbiciformis*, *P. personata* and *Xylota xanthocnema*. Studies using artificial rot-holes would also be beneficial, both as a survey technique, and to enhance the available habitat (see Biological Observations, page 36).

Management of ancient and decaying trees is a key issue and there is an urgent need to educate arboricultural officers, land agents and others concerned with management of woodlands and parkland trees about the value of different types of dead wood. Changing management practices is a key to maintaining the diversity of the saproxylic fauna of Surrey.

The Biodiversity Action Plan also addresses the problem of habitat management and, in this respect, Surrey is important for its heathland and chalk grassland, both of which need attention to reverse declines resulting from inadequate management over perhaps the last 40 years. There are other key habitats such as our ancient woodlands, and others which might be overlooked. For example, in a local context, base-rich flushes and seepages are very rare, so identification, safeguard and management of these sites are essential. One positive action would be to link future mineral planning consents to wetland habitat creation, so that new sites with reed beds or rich marginal vegetation are created. Other action might include allocation of visitor-free areas on parts of well-visited ponds in an attempt to reduce visitor pressure and restore areas rich in emergent vegetation and associated invertebrates.

Finally, current biodiversity planning relies on the environmental audit process, to which this atlas contributes. It will be necessary to repeat this survey in due course, to a similar degree of intensity, in order to assess changes, be they positive or negative. Detailed record-keeping by recorders is essential if that is to happen.

ACKNOWLEDGEMENTS

County mapping schemes cannot hope to achieve reasonable coverage without contributions from a large number of recorders and this scheme is no exception. I would therefore like to thank all the recorders listed below who have forwarded records over the years either directly to the Surrey scheme, via the .national scheme, as exhibitors of specimens at the British Entomological and Natural History Society annual exhibition, or as published records. The list of contributors, below, also includes those long-dead entomologists whose records have been gleaned from the literature or from collections now held in a variety of locations such as the Natural History Museum of London.

M. Aldridge
Dr K.N.A. Alexander (KNAA)
Sir C.H. Andrewes (CHA)
H.W. Andrews
J. Balfour-Browne
Dr S.G. Ball
F. Bancroft (FB)
S. Beck
P.L.T. Beuk (PLTB)
K.G. Blair (KGB)
A. Brackenbury
J.H. Bratton
E.J. Bunnett
P. Butcher
C.G. Champion (CGC)
P.J. Chandler (PJC)
P. Clarkson
D.J. Clark (DJC)
R.L. Coe (RLC)
J.H. Cole (JHC)
D.A. Coleman
J.E. Collin (JEC)
G.A. Collins (GAC)
C.N. Colyer (CNC)
F.J. Coulson
P.W. Currie (PWC)
H.V. Danks
A.S. Davidson
A.M. Davis
F.H. Day (FHD)
Dr J.S. Denton (JSD)
H.G. Denvil
Capt. C. Diver (CD)
J.R. Dobson (JRD)
R.D. Dunn (RDD)
M. Edwards (ME)
D. Element (DE)
Dr P.F. Entwistle
H.C. Eve (HCE)

S.J. Falk (SJF)
J.C. Felton
G. Fox-Wilson (GFW)
A.P. Fowles (APF)
N.D. Frankum (NDF)
R. Fry (RF)
D.J. Gibbs (DJG)
P. Goddard
A. Godfrey (AG)
C. Gorham
P. Grainger (PG)
A. Greensmith
S.J. Grove (SJG)
E.W. Groves
G.C.D. Grififths (GCDG)
K.M. Guichard (KMG)
A.J. Halstead (AJH)
C.O. Hammond (COH)
J.A. Hardman (JAH)
C. Hart
A. Harvey (AH)
P.R. Harvey (PRH)
R.B. Hastings (RBH)
C.N. Hawkins (CNH)
R.D. Hawkins (RDH)
P.J. Hodge (PJH)
V. Howard (VH)
Dr M.L. Howe (MAH)
Dr E. Howe (EAH)
M.O. Hughes (MOH)
Dr P.S. Hyman (PSH)
D. Iliff
S.F. Imber (SFI)
J.D. Ingham
F.B. Jennings (FBJ)
R.O. Jermyn (ROJ)
Ms C.M. Johnstone (CMJ)
A.W. Jones (AWJ)
R.A. Jones (RAJ)

L.N. Kidd
Dr P. Kirby
T.W. Kirkpatrick (TWK)
J. Kramer
H. Last
R. Leeke
D.C. Lees (DCL)
A. Le Gros
A.M. Low (AML)
G. Matthes (GM)
Dr I.F.G. McLean (IFGM)
A.V. Measday
R.K. Merrifield
S.R. Miles (SRM)
A.J.D. Morris (AJDM)
R.K.A. Morris (RKAM)
D.A. Moore (DAM)
G.E. Nixon (GEN)
Prof. J.A. Owen
L. Parmenter (LP)
M.S. Parsons (MSP)
R.M. Payne
C.W. Plant (CWP)
Dr J. Pontin (JP)
M.N. Pugh
O.W. Richards (OWR)
J.A. Riley
Mrs R. Robins
J. Roche (JR)
J.H.P. Sankey (JHPS)
Dr D.A. Sheppard (DAS)
P. Skidmore
M. Smart (MS)
K. Sorensen
W.H. Spreadbury (WHS)
B.M. Spooner
G.M. Spooner
J.B. Steer (JBS)
F.M. Struthers (FMS)

A.E. Stubbs (AES)   Rev. A. Thornley (AT)   L.K. Ward (LKW)
M. Sullivan   R.W.J. Uffen (RWJU)   P. Withers (PW)
M.C. Swan   G.H. Verrall (GHV)   L.S. Whicher
E.E. Symes   C.W. Wainwright (CWW)   A.J.A. Woodcock
D. Tagg (DT)   S. Wakely (SW)   J.W. Yerbury (JWY)

Surrey Biological Records Centre (SBRC)
Natural History Museum of London (NHML)

I am particularly grateful to Graham Collins who has accompanied me on many outings to record hoverflies; his contribution is considerable, especially as he has also endured my driving whilst ably navigating for much of the time. Roger Hawkins, another stalwart recorder, has supplied a substantial number of records, especially for south-east Surrey and for flower visit associations which are particularly valuable and enrich this account. I also thank Alan Stubbs for access to his duplicate of the card index compiled by the late Len Parmenter. Unless specifically stated, I have taken records to be those of Len Parmenter. I have also extracted the few Surrey records from the card index prepared by Steven Falk as part of the National Review of Flies (Falk, 1991). This included some data from museum collections which I have been unable to follow up.

Some 1,500 records submitted to the national Hoverfly Recording Scheme prior to 1990 have also been extracted; these include data held on the Invertebrate Site Register (ISR), set up by the Nature Conservancy Council, which now resides with the Institute of Terrestrial Ecology at Monks Wood. This helps to explain why some records by Alan Stubbs cover a date range because some of his data were abbreviated on the ISR.

I would also like to thank Alan Stubbs for much help and encouragement over the years, and not least for producing his guide to hoverflies which was the stimulus for my interest in these fascinating insects. Finally, I thank Dr Stuart Ball whose considerable patience and assistance with the computing aspects has made this project possible.

EXPLANATION OF SPECIES ACCOUNTS

These accounts are intended to interpret the data in the light of my experience in Surrey and the apparent trends that the maps exhibit. Detailed examination of the factors such as larval habits, drift geology and land use which might affect this distribution require consideration beyond the scope of this study, but they have been explored to some extent where possible. I have searched the literature for additional information relating to observations on hoverflies in Surrey and, although this has not been exhaustive, I hope that I have covered most of the obvious sources. Most relate to the bigger and more easily identified species, but comments on some of the scarcer, yet less obvious species are available. Unfortunately, this illustrates the tendency of recorders to concentrate on easily identified species and demonstrates the need to consider detailed observations on the commoner, yet less attractive species about which we know surprisingly little. Where appropriate, I have included notes on larval biology as a clue to the species' ecological preferences. These notes are based on published data gathered more widely than from Surrey, and on comments received on earlier versions of this text.

Chandler (*in press*) discusses the flux in current syrphid phylogeny. In his checklist the genera are arranged in alphabetic order, but many readers may not have access to this list and will probably use the names in Stubbs & Falk (1983). I have therefore followed the supra-generic structure used by Stubbs & Falk (*loc. cit.*), but have adopted the changes in generic and specific names accepted by Chandler (*loc. cit.*). Some of the changes were adopted in Stubbs (1996), but are also listed with the key changes listed below:

Old name	New name
Baccha	*B. obscuripennis* has been synonymised under *B. elongata*
Pyrophaena	***Platycheirus*** s.g. *Pyrophaena*
Dasysyrphus lunulatus	*D. **pinastri***
Doros conopseus	*D. **profuges***
Epistrophe melanostoma	Species new to Britain (Beuk, 1990)
Epistrophella euchroma	***Meligramma euchromum***
Melangyna guttata	***Meligramma guttatum***
Melangyna triangulifera	***Meligramma trianguliferum***
Megasyrphus annulipes	***Eriozona erratica***
Metasyrphus	***Eupeodes***
Sphaerophoria abbreviata	*S. **fatarum***
S. menthastri	*S. **interrupta***
Callicera aenea	*C. **aurata***

Old Name	New name
Cheilosia honesta	*C. **lasiopa***
C. intonsa	*C. **latifrons***
C. nasutula	*C. **vicina***
Chrysogaster chalybeata	*C. **cemiteriorum***
C. hirtella	***Melanogaster** hirtella*
C. macquarti	***Melanogaster aerosa***
Lejogaster splendida	*Lejogaster **tarsata***
Myolepta luteola	*M. **dubia***
Orthonevra splendens	***Riponnensia** splendens*
Eristalis nemorum	*Eristalis **interruptus***
Neocnemodon	***Heringia** s.g. Neocnemodon*
Pipizella varipes	*P. **viduata***
Microdon eggeri	*M. **analis***

THE MAPS

These show the distribution of each species with solid circles representing records from 1980 to 1997, open circles those prior to 1980 and, in one instance, grey circles which depict unconfirmed larval records. There are a number of instances where the recorder has provided a date range spanning the 1980 cut-off for the two date classes; in these cases, the mapping program treats the record as lying within the earlier date class and it appears as an open circle even though it is possible that the species has been recorded post-1980.

THE PHENOLOGY HISTOGRAMS

A small number of histograms have been included to depict the phenology of selected species and genera. Each month has been divided to cover the first 15 days and the second 15 or 16 days. Ideally, histograms would have been included for all species, but this would have meant greatly increasing the size and cost of this volume; they are consequently restricted to a few genera where there are noteworthy differences in emergence times, which may assist in separating similar species. For example, there appear to be differences in the emergence times of *Epistrophe melanostoma* and *E. nitidicollis*.

SPECIES STATUSES

National status

This account includes a national status where these have been published in Falk (1991). They comprise four classes:

Red Data Book (RDB) which is sub-divided into three categories representing the degree of perceived threat:

RDB 1 - Endangered
Taxa in danger of extinction.

RDB 2 - Vulnerable
Taxa believed likely to move into the Endangered category in the near future if the causal factors continue operating.

RDB 3 - Rare
Taxa with small populations that are not at present Endangered or Vulnerable, but are at risk.

Nationally Scarce formerly listed as Nationally Notable but changed to Nationally Scarce to maintain consistency with statuses used for plants.

Surrey Status

I have attempted to provide a real measure of each species' frequency in Surrey. This has been done by dividing the number of tetrads from which a species is recorded (N) by the number of tetrads for which there are records (R) and expressing the resulting percentage in a series of classes as outlined below:

Frequency = N/R x 100

No. of Tetrads	Frequency	Status
205+	40%+	Ubiquitous
77-204	15-40%	Common
21-76	4-15%	Local
11-20	2-4%	Scarce
1-10	>2%	Rare

NUMBER OF RECORDS

This simply identifies the number of records held on the database. With data coming from a number of sources it is likely that there will be some duplication, but this is extremely difficult to sort out. Consequently the true number of records may be marginally lower. For the purposes of the database I have considered a record to be that of a species recorded at a particular location on a particular day by one person. Where two or three recorders were present, their personal lists have been incorporated, thus duplicating records at a site on that day, but at the same time giving some additional measure of each species' frequency. Hence, in the records section it will be seen that on many occasions a record will be attributed to both Graham Collins and me, for example, and is counted as two records for the purpose of counting records.

FLIGHT TIMES

I have used the data for Surrey to give an indication of the broad span of times when each species flies. Peaks are indicated where these are apparent, and when a species is multiple-brooded the main peak is underlined, although in some cases the two peaks are about the same, so both are underlined.

CONSERVATION

In the event that obvious action can be taken to maintain and enhance existing populations, I have added notes on possible action which may benefit the species.

RECORDS

I have chosen to publish the records of hoverflies which are classed as Red Data Book or Nationally Scarce (Falk, 1991), or are scarce in Surrey and recorded in fewer than ten post-1980 tetrads, in order that a full account can be referred to in the future. This is unlikely to threaten the rarer species because most are infrequently seen and are unlikely to be caught in such numbers as to affect local populations. In fact, I believe that in the case of many species where conservation measures can be applied, there are positive advantages to having published records to which site managers can refer, and as a source of information for recorders who can supply information on the status of such species in the future. It is hoped that this study will provide a sound foundation for re-survey in years to come, and provides some indication of the possibilities for monitoring our more important species.

FLOWER VISITS

These relate solely to records of flower visits by hoverflies in Surrey. They mainly comprise my own records and those of Roger Hawkins, but also include those of other contributors and a variety of literature records – in the case of the latter, the author is quoted; most are by Len Parmenter and, to save space, I have abbreviated the reference to his records as LP. In some of the accounts, I have drawn attention to flower-visit preferences, some of which do not appear in the lists of flower visits. This is because these are recollections from field trips prior to my taking an interest in recording flower visits, and consequently these must be taken as anecdotal. In a few cases I can remember taking specimens from particular flowers and these have been quoted in the lists of flowers, even though they may not appear in my diaries. Plant names used in this text follow the nomenclature in Stace (1991).

Species accounts

SYRPHINAE

Bacchini

Larvae of this tribe are predatory, mainly on aphids but in at least one instance (*Xanthandrus comtus*) on lepidopterous larvae. Many are generalised aphid predators, but some are more specialised. Very little is known about the early stages of *Melanostoma* and many *Platycheirus*. Many of the adults are widely distributed and common, providing very few indications of particular habitat preferences.

Genus *Baccha*

These are narrow-bodied, elongate and highly distinctive flies. In Stubbs & Falk (1983) two species were listed, but there has always been some doubt about the differences between them. In Peter Chandler's new checklist (*in press*), the two *Baccha* species have been synonymised and this would seem to be the best treatment for them, as the separation between the two is both obscure and uncertain. However, this survey was conducted when the two separate species were recognised and as such there may be merit in illustrating their respective distributions.

Baccha spp.

Number of records: 134
Surrey Status: Common
Flight times: April – November
Peaks: <u>May</u>, August

This map comprises records for unidentified *Baccha* specimens, mainly females but also individuals which were not captured, and the records for the formerly separate *B. elongata* and *B. obscuripennis*. There do not appear to be discernible trends for this species which inhabits low-growing scrubby vegetation in woodland rides and at woodland edge; this is consistent with the recorded larval biology which is reported by Rotheray (1993) as "associated with . . . ground layer aphids in shaded sites".

Flower visits: cow parsley, pignut, black bryony

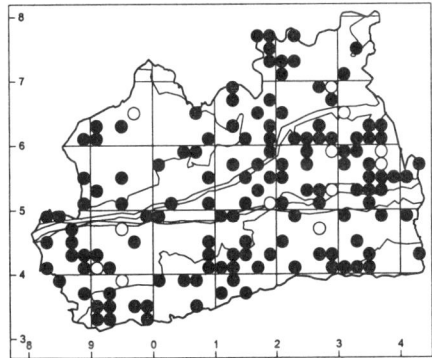

Baccha elongata (Fabricius, 1775)

Number of records: 20

Flight times: May – August

Peaks: <u>May</u>, August

If this is a valid species, *B. elongata* appears to be much less common than *B. obscuripennis*.

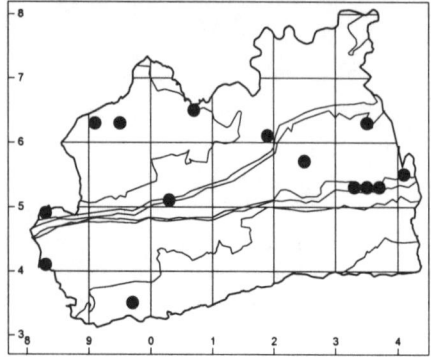

Baccha obscuripennis Meigen, 1822

Number of records: 67

Flight times: April – November

Peaks: <u>May</u>, July/August

This appears to be a widely distributed species, if it is truly separate from *B. elongata*.

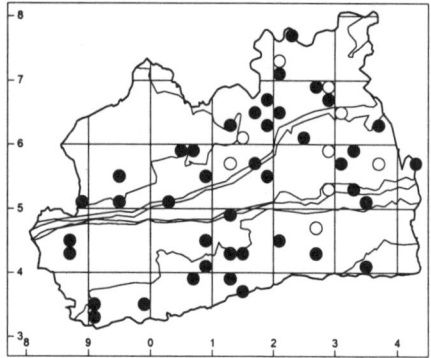

Genus *Melanostoma*

These are small narrow species which show distinct sexual dimorphism. The males have rectangular yellow abdominal spots and the females have triangular abdominal spots. The larvae are poorly known and although the adults are common in grassland and at woodland edge, larvae are rarely found.

Melanostoma mellinum (Linnaeus, 1758)

Number of records: 458

Surrey Status: Ubiquitous

Flight times: April – October

Peaks: <u>May</u>, August

This common species is most frequently swept from grassland or found at plantain flowers. Although the maps suggest superficially that this species is as widespread as *M. scalare*, there are far fewer records. Analysis of detailed records collected between 1994 and 1996 indicates that *M. mellinum* is considerably scarcer than *M. scalare* (61 records against 213 records) so, although it may be overlooked to some extent, there is a real difference in frequency. There is one record of this species as a prey item of the fly *Empis tessellata* (Diptera, Empididae) (LP, 1968).

Flower visits: marsh marigold, creeping buttercup (LP, 1950a), buttercup spp., lesser spearwort (LP, 1950a), greater stitchwort (LP, 1950a), cuckoo-flower (LP, 1950a), bastard cabbage, heather, hawthorn, tormentil (LP, 1950a), rowan, ivy, wood spurge (LP, 1957), wild angelica, hogweed, field bindweed, ribwort plantain, nipplewort (LP, 1950a), perennial sow-thistle, prickly sow-thistle (LP, 1950a), bluebell (LP, 1950a)

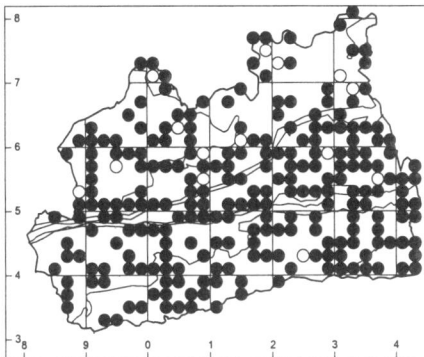

Melanostoma scalare (Fabricius, 1794)

Number of records: 903

Surrey Status: Ubiquitous

Flight times: April – September

Peaks: <u>May</u>, August

This common grassland hoverfly is frequently found at grass and plantain flowers, especially along woodland rides. It is sometimes susceptible to fungal attack, especially in the autumn when dead specimens may be found in numbers attached to seed heads of umbellifers and plantains, together with *M. mellinum*.

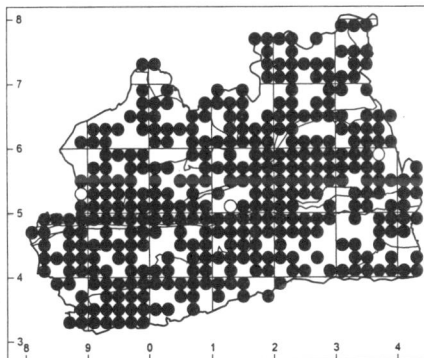

Flower visits: creeping buttercup, buttercup spp., lesser celandine, greater stitchwort, lesser stitchwort, water chickweed, broad-leaved dock, garlic mustard, heather, bramble, wood avens, blackthorn, wild cherry, hawthorn (LP, 1957), rosebay willowherb, dogwood, dog's mercury, ivy, cow parsley, wild angelica (Uffen, 1969), wild parsnip, hogweed, upright hedge-parsley, field bindweed, hedge bindweed, water mint (LP, 1950a), ribwort plantain, honeysuckle, devil's-bit scabious, creeping thistle, cat's-ear, dandelion, common fleabane, goldenrod, Canadian goldenrod, common ragwort, ragwort spp., bluebell (LP, 1955b)

Genus *Platycheirus*

This is a very variable group of narrow-bodied species. There is obvious sexual dimorphism in many species, and the modified male front legs are a valuable clue to their identity. The larvae of some are highly prey-specific, for example *P. fulviventris*, whilst others are polyphagous. The former genus *Pyrophaena* has been subsumed into this genus at sub-generic level.

Platycheirus ambiguus (Fallén, 1817)

Number of records: 30
Surrey Status: Local
Flight times: March – May
Peak: April

Although the map suggests that this species is most frequent in the London area, it is likely that it will prove to be more widespread. This is because *P. ambiguus* is a spring species which frequents blackthorn blossom amongst others, and opportunities to examine blackthorn in good weather are often limited; the author lives in south London and there is therefore some recorder bias as short local trips are prevalent in the spring.

Flower visits: blackthorn

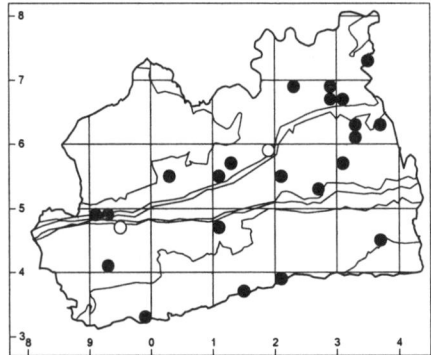

Platycheirus albimanus (Fabricius, 1781)

Number of records: 927
Surrey Status: Ubiquitous
Flight times: March – October
Peaks: <u>May</u>, August

This is the commonest species of *Platycheirus* which occurs in a variety of habitats, especially woodland edge and hedgerows, and peaks in spring and in late summer. *P. albimanus* is often in such abundance that it is possible that similar species such as *P. ambiguus* and *P. sticticus* are overlooked. Once the separation from *P. discimanus* is known, females are readily identified in the field, but the males, which are variable in colouration, are best identified using a lens.

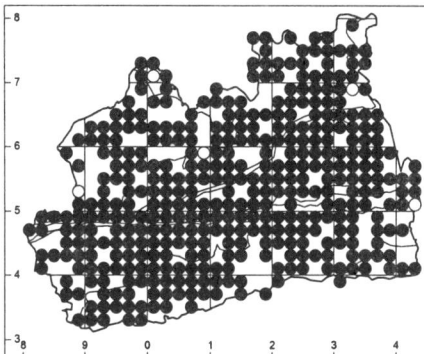

Flower visits: marsh marigold, creeping buttercup, buttercup spp., water crowfoot (LP, 1950a), lesser celandine, chickweed, greater stitchwort, lesser stitchwort, water chickweed, hedge mustard (LP, 1950a), garlic mustard, cuckoo-flower (LP, 1950a), heather, *Erica* spp., primrose (LP, 1950a), tormentil, blackthorn, wild cherry, rowan, hawthorn, black medick, square-stalked willowherb, herb robert, ivy, wood spurge (LP, 1957), cow parsley, burnet-saxifrage, hemlock water-dropwort, wild angelica (Uffen, 1969), wild parsnip, hogweed, upright hedge-parsley, field bindweed, water forget-me-not (LP, 1950a), white dead-nettle, bugle (LP, 1950a), ground-ivy, wild basil, water mint, *Hebe* sp., red bartsia, welted thistle, nipplewort, cat's-ear, autumn hawkbit, rough hawkbit, perennial sow-thistle, prickly sow-thistle (LP, 1950a), dandelion, common fleabane, scentless mayweed, ragwort spp., water-plantain, ramsons, bluebell (LP, 1950a)

Platycheirus angustatus (Zetterstedt, 1843)

Number of records: 57
Surrey Status: Local
Flight times: April – September
Peaks: <u>May</u>, August

One of the more difficult species to identify in the original key in Stubbs & Falk (1983) and therefore possibly overlooked, *P. angustatus* is now relatively simple to distinguish from *P. clypeatus* (Stubbs, 1996). Males are generally smaller, thinner and darker than those of *P. clypeatus* and females have narrower bodies and a pointed abdomen. Even so, it is necessary to retain material for microscopic examination, as this is a species which cannot be identified in the field with any confidence. This is a grassland hoverfly

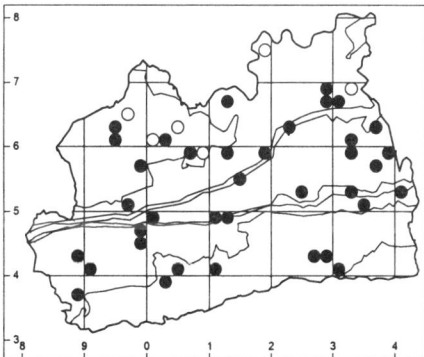

which is more frequent in damp rushy pasture and woodland rides and appears to be much scarcer on the Chalk where drier conditions prevail.

Flower visits: heather, upright hedge-parsley, gipsywort (LP, 1950a), sneezewort (LP, 1950a), dandelion

Platycheirus clypeatus agg.

Number of records: 151

Platycheirus clypeatus has recently been shown to be a complex of four species (Goeldlin de Tiefenau, Maibach & Speight, 1990), of which only two (*P. clypeatus* s.s. and *P. occultus*) have been recorded from Surrey. This means that all previous records of *P. clypeatus* must be aggregated and only records confirmed through voucher material allocated to the new species. This map merely serves as a reminder to all recorders to retain voucher material as there are possibilities of further splits among many of the more difficult species. Although only *P. clypeatus* s.s. and *P. occultus* have been recorded in Surrey, the possibility of *P. europaeus* should not be discounted as suitable localities may occur, for example, in the clay woodlands of the Low Weald.

Flower visits: buttercup spp., garlic mustard

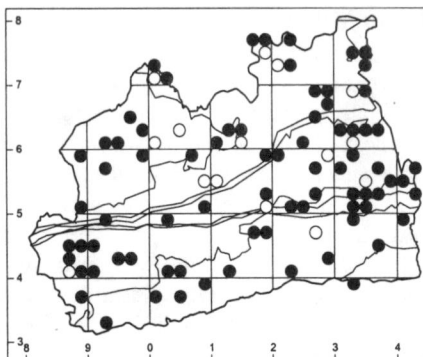

Platycheirus clypeatus s.s.

(Meigen, 1822)

Number of records: 90

Surrey Status: Common

Flight times: April – September

Peaks: May, <u>August</u>

This grassland species is double-brooded but considerably more common in late summer (see *Figure 4*). *P. clypeatus* appears to be the most ubiquitous member of the "*clypeatus*" species complex. Given the overall frequency of *P. clypeatus* agg., most of whose records probably refer to *P. clypeatus* s.s., I have adjusted the status for this species from "Local" to "Common".

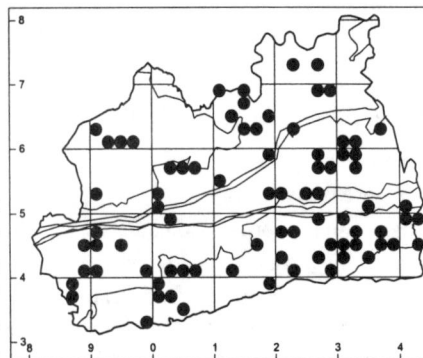

Flower visits: redshank, black medick, wild parsnip

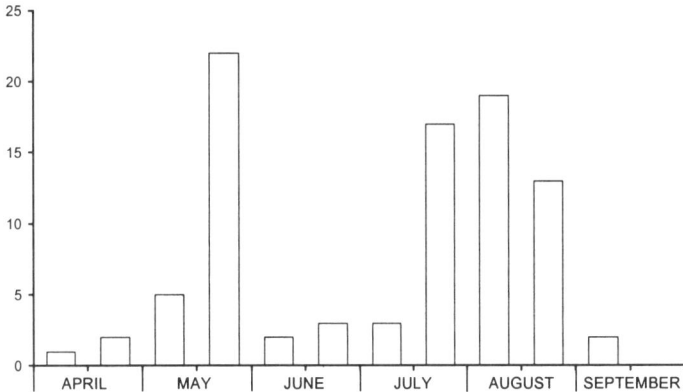

Figure 4. Phenology of *Platycheirus clypeatus*

Platycheirus discimanus Loew, 1871

Nationally Scarce
Number of records: 8
Surrey Status: Rare
Flight times: April – June
Peak: April/May

One of the early spring species which visit sallow catkins and blackthorn blossom, *P. discimanus* appears to be extremely scarce in Surrey. There are very few conclusions to be drawn from the data other than it is apparently a woodland hoverfly which possibly favours older or ancient woods. It also appears to have

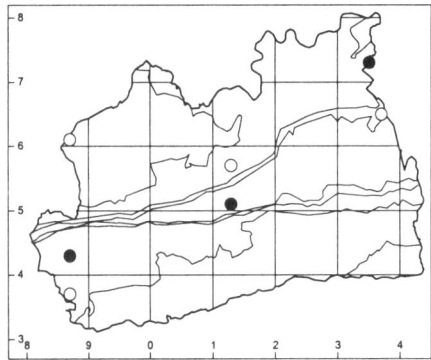

undergone a significant national decline (see Ball & Morris, *in press*).

Flower visits: sallow, blackthorn

RECORDS: **Camberley** (10.6.1926, OWR); **Hindhead** (6.5.1922, labelled R.J. in Doncaster Museum); **Tilford Reeds** SU8643 (3.4.1988, RKAM); **Ranmore** TQ1350 (29.4.1987, GAC); **Great Bookham Common** (no date, LP); **Sydenham Hill Woods** TQ3472 (14.4.1988, AG); **Shirley** (5.5.1872, 15.5.1875, GHV).

Platycheirus fulviventris (Macquart, 1829)

Number of records: 19

Surrey Status: Rare

Flight times: May – September

Peak: June

This wetland species is usually found by pond margins and in wet meadows. It is a brightly-coloured hoverfly which is often seen flying through tall vegetation rather than visiting flowers. It is met with infrequently in Surrey where ponds rich in marginal vegetation are scarce.

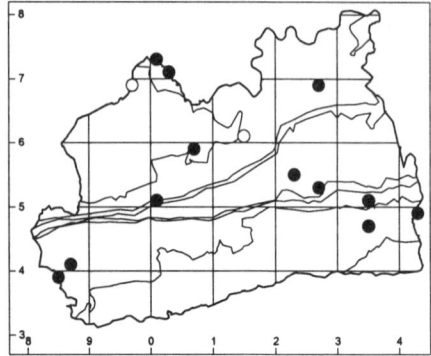

Platycheirus immarginatus (Zetterstedt, 1849)

Nationally Scarce

Number of records: 1

Surrey Status: Extinct

Verrall (1901) reports catching this species "in very considerable numbers by the side of the Thames near Kew on July 16th 1868". Females are difficult to separate from *P. clypeatus* and so the record must be treated with caution, but given its provenance I have accepted it. At that time, vegetated brackish river margins may have been more frequent than today and it is likely that the extinction of this species was due to containment of the tidal river Thames within flood defences, with the consequent loss of inter-tidal marsh.

Platycheirus manicatus (Meigen, 1822)   PLATE 8

Number of records: 55

Surrey Status: Local

Flight times: April – September

Peaks: <u>May</u>, August

Stubbs & Falk (1983) describe *P. manicatus* as a dry grassland species which occurs on alkaline or neutral soils; this is borne out by the known distribution in Surrey which is concentrated on the Chalk. Rotheray (1993) reports, however, that gravid females are associated with moist shady sites. These observations seem to be contradictory, especially as *P. manicatus* is often abundant in coastal locations with little or no cover. The picture is further complicated because there seem to be close similarities between the distribution of *P. manicatus* and that of *P. tarsalis*, although the latter is seemingly more widely distributed and closely associated with woodland. These conflicting observations

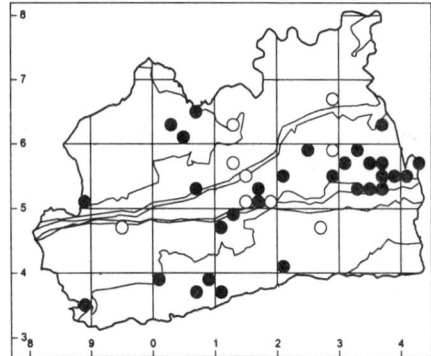

make interpretation of the map very difficult and I cannot offer an explanation for the distribution as it stands. There is one record of this species suffering fungal attack (RKAM).

Flower visits: greater stitchwort (LP, 1950a), yellow pimpernel (LP, 1950a)

Platycheirus occultus Goeldlin de Tiefenau, Maibach & Speight, 1990

Number of records: 10
Surrey Status: Rare
Flight times: April – August
Peak: June

One of only two species of the *"clypeatus"* complex to have been found in Surrey, *P. occultus* is clearly associated with acidic habitats such as those found on heathland. It is a good indicator of wet heathland in Surrey, although it seems to favour peaty habitats on rich and poor fen elsewhere in Britain (S. Falk, *pers. comm.*). Males are readily identified on

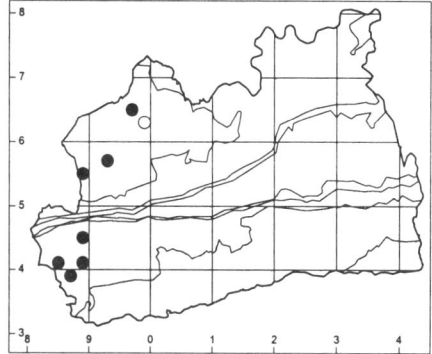

front tarsal characters, but the females can be more problematic. A good voucher collection is therefore essential. Although the range of dates for this species suggests that its flight period is similar to that of *P. clypeatus* (see *Figure 4*), there are insufficient records to be sure of any pattern (with no more than two records for any two-week period). On a monthly basis, however, this species appears to peak in June.

RECORDS: **Frensham Little Pond** SU8541 (25.4.1987, RKAM); **Devil's Jumps** SU8639 (5.8.1989, RKAM); **Hankley Common** SU8940 (1.7.1989, RKAM); **Thundry Meadows** SU8944 (11.5.1991,13.7.1991, JRD); **Mytchett Lake** SU8954 (18.6.1989, RKAM); **Furze Hill Pool** SU9356 (18.6.1989, RKAM); **Chobham Common** SU9665 (3.6.1979, SJF); **Chobham Longcross** SU9765 (10.5.1987, RKAM); **Chobham Common, Gracious Pond** SU9963 (3.6.1979, SJF/AES)

Platycheirus peltatus agg.

Number of records: 80
Flight times: April – September
Peaks: <u>May</u>, August

This is another species complex which has recently been split, but unlike *P. clypeatus* the separation appears to follow a north/south divide with *P. nielseni* mainly occurring in the north and *P. peltatus* in the south. This map illustrates the distribution of records for which there are no vouchers, and again serves to illustrate the need to retain voucher material.

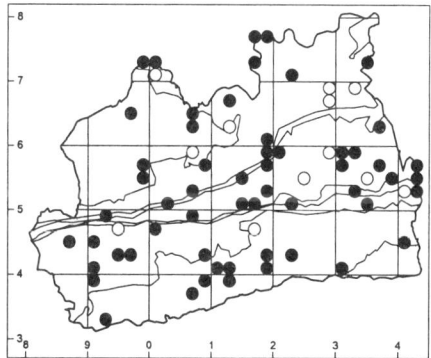

Platycheirus peltatus s.s. (Meigen, 1822)

Number of records: 22

Surrey Status: Local

Flight times: May – September

Peaks: <u>May</u>, August

This species is usually associated with rank grassland, but there would appear to be no obvious pattern to its distribution. There would appear to be substantial differences in the frequency of the species over the years, as illustrated in *Figure 5*. Moreover, this is one of a number of species which has been markedly scarcer in the 1990s, as is clear from the table. The principal reason for this change is likely to be drought, the effects of which are clearly illustrated by the slump in numbers recorded since the first major drought in 1990/1991.

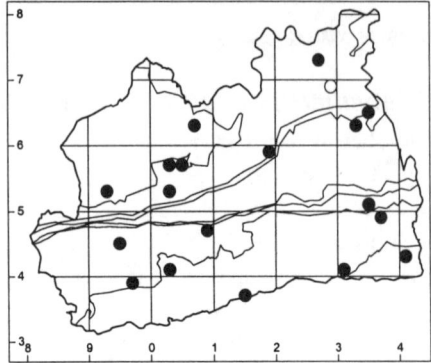

Flower visits: buttercup spp., square-stalked willowherb, water mint

	85	86	87	88	89	90	91	92	93	94	95	96
P. peltatus agg.	4	13	31	13	12	5	2					
P. peltatus s.s.	4	6							2	6	3	1
Total	8	19	31	13	12	5	2	0	2	6	3	1
% of records	0.4	1	0.8	0.5	0.45	0.5	0.2	0	0.1	0.2	0.15	0.1

Figure 5. **Numbers of *Platycheirus* agg. and *P. peltatus* s.s. recorded from 1985 to 1996. This demonstrates the yearly fluctuation in numbers and an apparent decline since 1990.**

[*Platycheirus scambus* (Staeger, 1843)

The single old record for Great Bookham Common (LP, 1966) needs to be treated with caution as *P. scambus* is essentially a northern species. There have, however, been a small number of confirmed records from southern England in recent years and this record cannot be completely discounted.]

Platycheirus scutatus (Meigen, 1822)

Number of records: 303
Surrey Status: Common
Flight times: April – October
Peaks: <u>May</u>, August

This is one of the commonest spring hoverflies which mainly occurs along woodland edge. It is multiple-brooded through the year. Although widely recorded, it is likely to be overlooked when *Platycheirus* spp. are common, as it can be rather insignificant and unassuming. There would appear to be no definite pattern to its distribution.

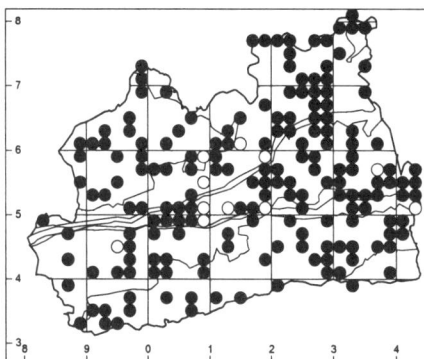

Flower visits: common nettle, chickweed, water chickweed, redshank, knotgrass, Japanese knotweed, garlic mustard, perennial wall-rocket, enchanter's nightshade, cow parsley, wild angelica (Uffen, 1969), hogweed, upright hedge-parsley, water mint, cat's-ear, perennial sow-thistle, smooth sow-thistle, dandelion, common fleabane, scentless mayweed, common ragwort, water-plantain, bluebell (LP, 1955b), ramsons

Platycheirus sticticus (Meigen, 1822)

Nationally Scarce
Number of records: 2
Surrey Status: Rare
Flight times: May – August

There are two pre-1980 records for this species which is small and inconspicuous, and possibly overlooked. Given its scarcity both in Surrey and nationally, it is not possible to identify any particular habitat associations.

RECORDS: **Box Hill** TQ1851 (26.5.1973, AES); **Earlswood Common** TQ2747 (1.8.1957, JHC).

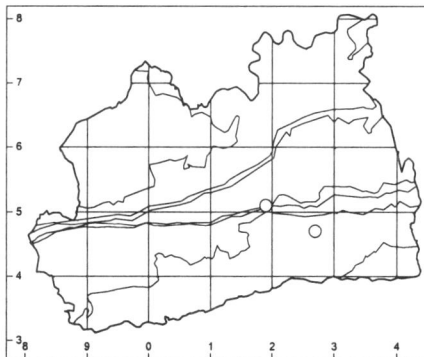

Platycheirus tarsalis (Schummel, 1837)

Number of records: 62
Surrey Status: Local
Flight times: April – August
Peak: May

Platycheirus tarsalis is a spring species which is most frequently found at garlic mustard in woodland rides and hedgerows. It is principally a woodland species and is commonest in clay woodlands. It has a curiously clumped distribution but is most frequent on the clay-with-flints overlying the Chalk in the east. This may be a true representation of its distribution, but may alternatively reflect the species' short flight period and corresponding recorder bias. This species is recorded as a prey item of *Empis tessellata* (Diptera, Empididae) by Parmenter (1966) who also records that a female laid a phenomenal 346 eggs whilst in captivity (LP, 1951); this seems to be one of the few accounts of the egg-laying capacity of hoverflies.

Flower visits: greater stitchwort, garlic mustard, herb robert, cow parsley, wayfaring tree (LP, 1941), dandelion, bluebell (LP, 1955b)

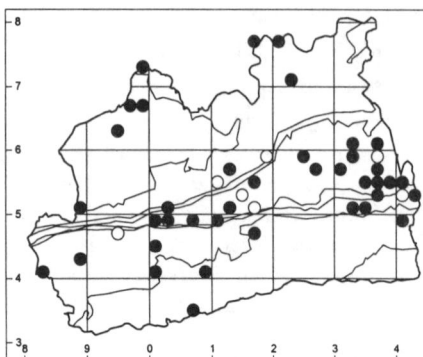

Platycheirus granditarsus (Forster, 1771)

Number of records: 144
Surrey Status: Common
Flight times: April – September
Peaks: May, <u>August</u>

This widespread species is associated with damp habitats such as pond margins, ditches and damp rushy pasture, but is sufficiently ubiquitous to suggest that its requirements are quite catholic. There is a clear concentration of records in south-east and north-west Surrey with far fewer records from the Chalk. Occasionally it occurs in considerable numbers, for example at Blindley Heath pond where many tens of individuals were seen in a short period on 10.8.1986.

Flower visits: lesser spearwort, goosefoot spp., redshank, silverweed (LP, 1950a), water-cress, wild parsnip, hogweed, common fleabane, scentless mayweed, water-plantain

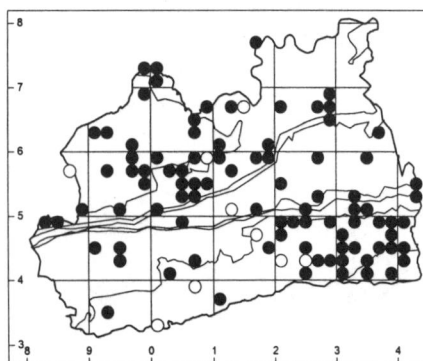

Platycheirus rosarum (Fabricius, 1787)

Number of records: 175

Surrey Status: Common

Flight times: May – September

Peaks: <u>June</u>, August

Platycheirus rosarum is widespread and associated with wet habitats, and is often found in association with *P. granditarsus.* There is a similar bias in distribution, but *P. rosarum* appears to exhibit stronger affinities with the London Clay and Low Weald. It is more frequently found flying amongst vegetation than visiting flowers.

Flower visits: creeping buttercup (LP, 1950a), goosefoot spp., redshank, silverweed (LP, 1950a), wild angelica, wild parsnip, hogweed, wild carrot, water forget-me-not (LP, 1950a)

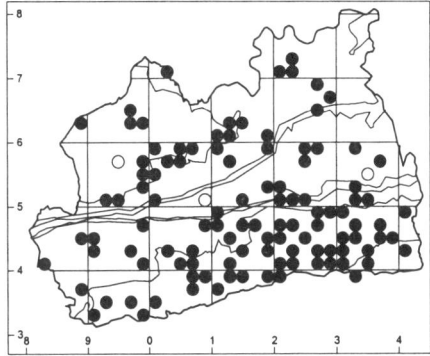

Genus *Xanthandrus*

Xanthandrus comtus (Harris, 1780)

Nationally Scarce

Number of records: 9

Surrey Status: Rare

Flight times: June – September

Peaks: July, <u>September</u>

There would appear to have been a brief surge in records of this striking hoverfly in recent years elsewhere in England, but in Surrey this has not been the case. Most records are of single individuals, but there is one record of many seen on heather on Chobham Common (Moore, 1989). Given its sporadic occurrence it could well be mainly migratory. The larvae are predatory upon gregarious caterpillars of micro-moths (Rotheray, 1993) and it is well worth examining such aggregations of larvae for the larva of *X. comtus.*

Flower visits: heather (Moore, 1989)

RECORDS: **Farnham** SU8449 (7.1994, in house, DT); **Chobham Common** SU9764 (8.7.1988, DAM); **Wisley RHS Gardens** TQ0658 (1982, AJH); **Ashtead Common** (23.7.1931, RLC); **Nonsuch Park** TQ2263 (11.9.1985, VH); **Banstead Heath** TQ2454 (21.9.1985, RKAM); **Richmond Park** TQ2072 (28.8.1992, MSP); **Selsdon** (1941, RLC); **Thornton Heath** (30.6.1940, LP).

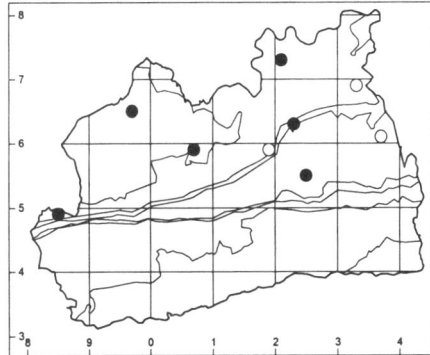

Paragini

Genus *Paragus*

There are three species in the British fauna, all of which have been recorded in Surrey, but only one is widespread. The larvae feed on a range of arboreal and ground-layer aphids (Rotheray, 1993).

Paragus haemorrhous Meigen, 1822

Number of records: 65
Surrey Status: Local
Flight times: May – September
Peaks: June, August

This widely distributed species is likely to have been overlooked because of its small size. Only the males could be identified until recently, and this means that a proportion of likely records have been ignored. *P. haemorrhous* is a grassland species which often frequents short grass at the edges of footpaths and disturbed sparsely-vegetated sites. Most records originate from the Chalk and the heathlands of west Surrey where hotter, drier conditions prevail. Two records from the Low Weald are surprising, but this may reflect the local drift geology. A female which was likely to have been this species was observed ovipositing on a cat's-ear rosette on Mitcham Common, TQ2867 (RKAM).

Flower visits: common rock-rose, germander speedwell, mouse-ear hawkweed

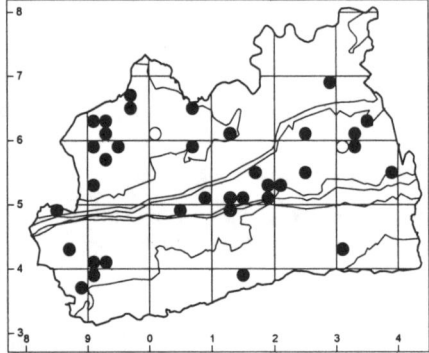

Paragus tibialis (Fallén, 1817)

Nationally Scarce
Number of records: 4
Surrey Status: Rare
Flight times: June – August
Peak: June

There are very few records of *P. tibialis* which is a known heathland associate (Stubbs & Falk, 1983). This is possibly because heathlands are often unproductive during its flight period in midsummer and may be under-worked as a result. Further detailed efforts to record this species may show that it is in fact more widespread than the records suggest.

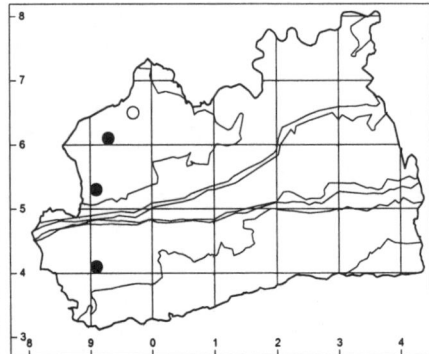

RECORDS: **Thursley Common** SU9041 (28.7.1991, JRD); **Ash Ranges - Fox Hills** SU9152 (13.8.1989, SJF); **Folly Bog** SU9261 (24.6.1997, JSD); **Chobham Common** SU9665 (3.6.1979, SJF).

Paragus albifrons (Fallén, 1817)

Red Data Book 2
Number of records: 1
Surrey Status: Extinct
There is only one record of this hoverfly, a specimen from Guildford in the Verrall collection at Oxford (SJF record cards). This is a species which seems to have become considerably scarcer nationally (Ball & Morris, *in press*) and may no longer be resident in Surrey.

Syrphini

These are amongst the most colourful species, many of which are wasp mimics. This is a group whose larvae are predators on aphids and, in one instance, beetle larvae. Some (*Chrysotoxum*) are thought to have associations with ants, and others (*Xanthogramma*) have been bred from ant nests. There are some obvious habitat associations such as those with broadleaved or coniferous woodland, woodland edge with umbellifers, heathlands and grasslands. Oddly, none seem to be closely associated with wetlands. Many species are ubiquitous or so widely distributed as to be presumed ubiquitous.

Genus *Chrysotoxum*

These are brightly-coloured wasp mimics which include a number of species which are difficult to separate. Retention of vouchers of *C. elegans*, *C. octomaculatum* and *C. verralli* is essential as they can be particularly troublesome to new recorders and include two nationally scarce species. The larvae are poorly known, but are thought to be associated with ant nests.

[*Chrysotoxum arcuatum* (Linnaeus, 1758)

Number of records: 3
Surrey Status: Unconfirmed – Doubtful
Verrall (1901) reports that he believed that he had "seen a specimen taken by Mr Saunders at Chobham in Surrey" and goes on to speculate that it may have been *C. octomaculatum*. It seems unlikely that *C. arcuatum* was a correct identification given the northern and western distribution of this species (Ball & Morris, *in press*); *C. octomaculatum* had not been recognised as separate from *C. verralli* at this time, so the exact identity of this specimen is not clear. There is also a record of this species on the national database from Godstone (FHD) in July 1952 which is extremely doubtful, and a further ancient record from Oxshott (Billups, 1891) which must be discounted for similar reasons.]

Chrysotoxum bicinctum (Linnaeus, 1758)

Number of records: 295
Surrey Status: Common
Flight times: May – September
Peak: August

This common hoverfly occurs mainly in grassland and scrubby habitats, but also in woodland rides. There would appear to be no clearly defined distribution to this species, although it seems to be less common in the London area where un-manicured open space is at a premium.

Flower visits: meadow buttercup, creeping buttercup, buttercup spp., lesser spearwort, lesser stitchwort, tormentil (LP, 1950a), common rock-rose, charlock, bramble, burnet-saxifrage, wild angelica, wild parsnip, hogweed, upright hedge-parsley, wild carrot, field bindweed, water mint, common marsh bedstraw, creeping thistle, common knapweed, autumn hawkbit, hawkweed oxtongue, mouse-ear hawkweed, common fleabane, yarrow, oxeye daisy

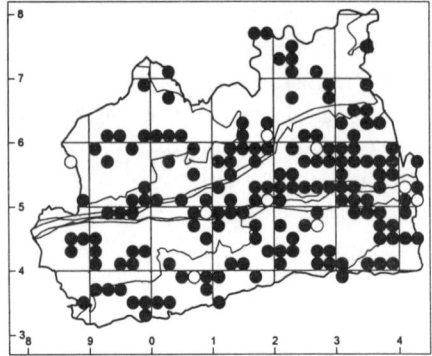

Chrysotoxum cautum (Harris, 1776)   PLATE 9

Number of records: 139
Surrey Status: Common
Flight times: May – July
Peak: June

Chrysotoxum cautum is widely distributed in Surrey, but appears less frequently in the Low Weald and in the London area. It is mainly a grassland species that is usually found in very low numbers and, although it visits flowers, it is more frequently seen in flight when it can show a striking resemblance to a worker social wasp. There is a record of a number of females "ovipositing at random on the grass and other vegetation in a sunny hedge bottom" at Eashing (Uffen, 1961).

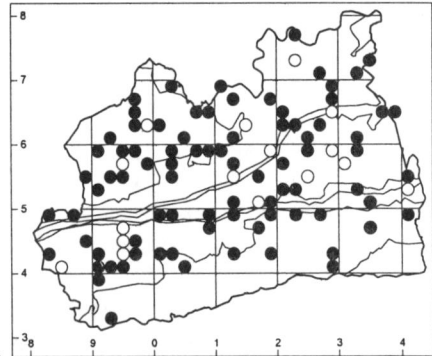

Flower visits: buttercup spp., dogwood, cow parsley, hogweed

Chrysotoxum elegans Loew, 1841

Red Data Book 3
Number of records: 19
Surrey Status: Rare
Flight times: April – August
Peaks: June, <u>August</u>

Superficially *C. elegans* resembles *C. cautum* in the field, but is slightly narrower in build; such specimens merit retention for confirmation; moreover it generally occurs later in the year (see *Figure 6*). *C. elegans* is a grassland species; most of the records are from the Chalk, but Ashtead and Bookham Commons are on clay; factors behind its apparent rarity are not clear. It would appear to be most frequently seen in western England (Ball & Morris, *in press*) and may be at the extreme edge of its distribution in Surrey. Just occasionally it can occur in large numbers, as at Box Hill where it has been seen at wild parsnip flowers in August. The specimen from Banstead Downs is the first post-1970 record for the London area (Plant, 1986).

Flower visits: wild parsnip

RECORDS: **Woking**, (pre-1940, JEC), (5.1886, GHV); **Gomshall** (27.8.1966, AES); **Hackhurst Downs** (24.5.1956, 5.1958, COH); **Great Bookham Common** (12.6.1948, COH), (30.8.1964, AWJ); **White Downs** (12.7.1947, [Wakely, 1949]); **Ranmore** TQ1250 (16.8.1987, RBH); **Box Hill** TQ1751 (4.8.1960, 9.8.1963, SFI), (21.8.1977, 26.4.1979, AES), (2.8.1982, DJG), (14.8.1987, RKAM), TQ1852 (28.7.1997, KNAA); **Ashtead Common** (26.5.1973, [Evans, 1974]); **Banstead Downs** TQ2460 (29.8.1985, RKAM); **Selsdon Wood** TQ3661 (7.8.1934, RLC).

Chrysotoxum festivum (Linnaeus, 1758)   PLATE 9

Number of records: 158
Surrey Status: Local
Flight times: June – September
Peak: August

There would appear to be a bias in distribution towards north and east Surrey where *C. festivum* can often be seen in grassland, scrubby localities and suburban gardens. This species seems to be commonest in warm dry habitats and it would appear to be scarce or absent on the clays of the Low Weald. Males are frequently found hovering in sunny places and there is a record of males resting on *Prunus* leaves and darting out as if to investigate passing insects, "possibly awaiting females" (Jones, 1995).

Flower visits: meadow buttercup, common rock-rose, water-cress, weld, bramble, agrimony,

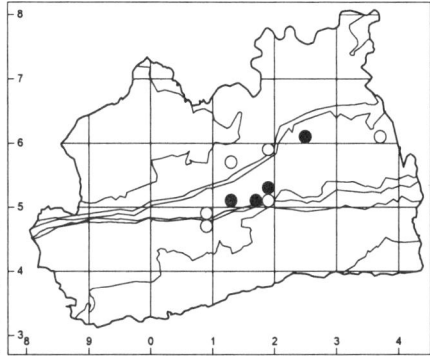

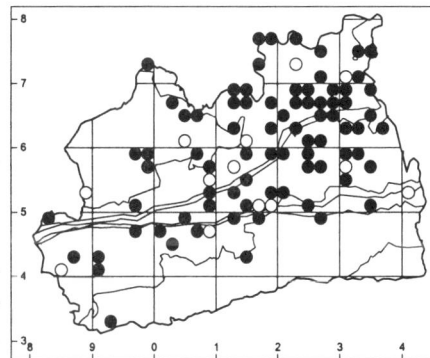

burnet-saxifrage, wild angelica, upright hedge-parsley, wild carrot, field bindweed, wild marjoram, water mint, ribwort plantain, eyebright, common knapweed, smooth hawk's-beard, Michaelmas-daisy, hoary ragwort

Chrysotoxum octomaculatum Curtis, 1837

Red Data Book 2
Number of records: 4
Surrey Status: Rare
Flight times: May – August
Peaks: May, August

Chrysotoxum octomaculatum has only been recorded from the heathlands of south-west Surrey, although there is a very doubtful record from Limpsfield Common (LP, 1942) and another from Oxted (Kirkpatrick, 1918). The account in Verrall (1901) which alluded to an earlier record under *C. arcuatum* (see species account, page 63) might suggest that this species occurred on Chobham Common at one time, but this observation was well before the differences between this species and *C. verralli* were recognised and must be discounted. Identification of *C. octomaculatum* is difficult and it is necessary that voucher specimens are retained with details of the precise habitat from which they were recorded; this is essential if this species' biology is to be properly understood and conservation measures are to be devised to reinforce the population. Although its biology is not known, *C. octomaculatum* is likely to be associated with ants in some way. Stephen Miles (*pers. comm.*) reports that his specimen was taken in a non-heathy area with buttercups, suggesting that this may be a species of heathland edges.

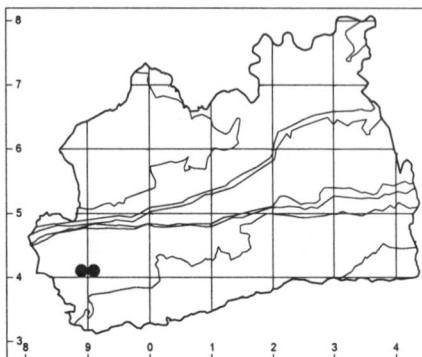

Conservation: *C. octomaculatum* is listed on the short list of the UK Biodiversity Action Plan (DoE, 1995), making the two localities from which it has recently been recorded particularly important. There are no obvious conservation measures that can be undertaken at this stage other than maintenance of existing *Calluna* heathland, including scrub edge and other habitats marginal to heathland.

Flower visits: buttercup spp. (Miles, 1989)

RECORDS: **Hankley Common** SU8840 (28.5.1988, SRM), (25.8.1992, JSD); **Thursley Common** SU9040 (28.5.1966, AES), (14.8.1989, JSD).

Chrysotoxum verralli Collin, 1940

Number of records: 65

Surrey Status: Local

Flight times: June – September

Peak: July

Many early records of *C. verralli* are likely to have been erroneously recorded as *C. octomaculatum* from which it was subsequently split. *C. verralli* typically occurs later than similar species and appears to be more frequent in north-east Surrey than elsewhere. It is mainly associated with lightly scrubbed grasslands and often visits such umbellifers as wild parsnip.

Flower visits: wild parsnip, hogweed, wild marjoram

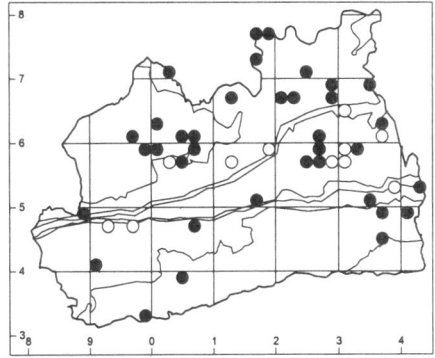

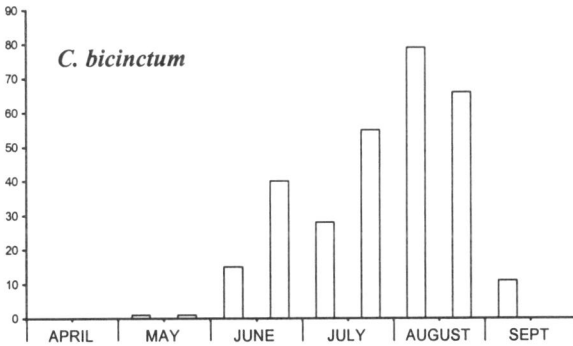

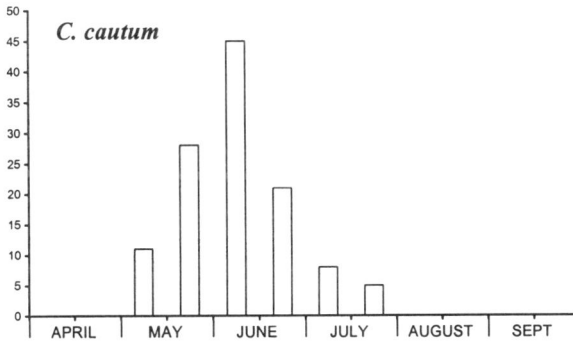

Figure 6a. Phenology of *Chrysotoxum* species

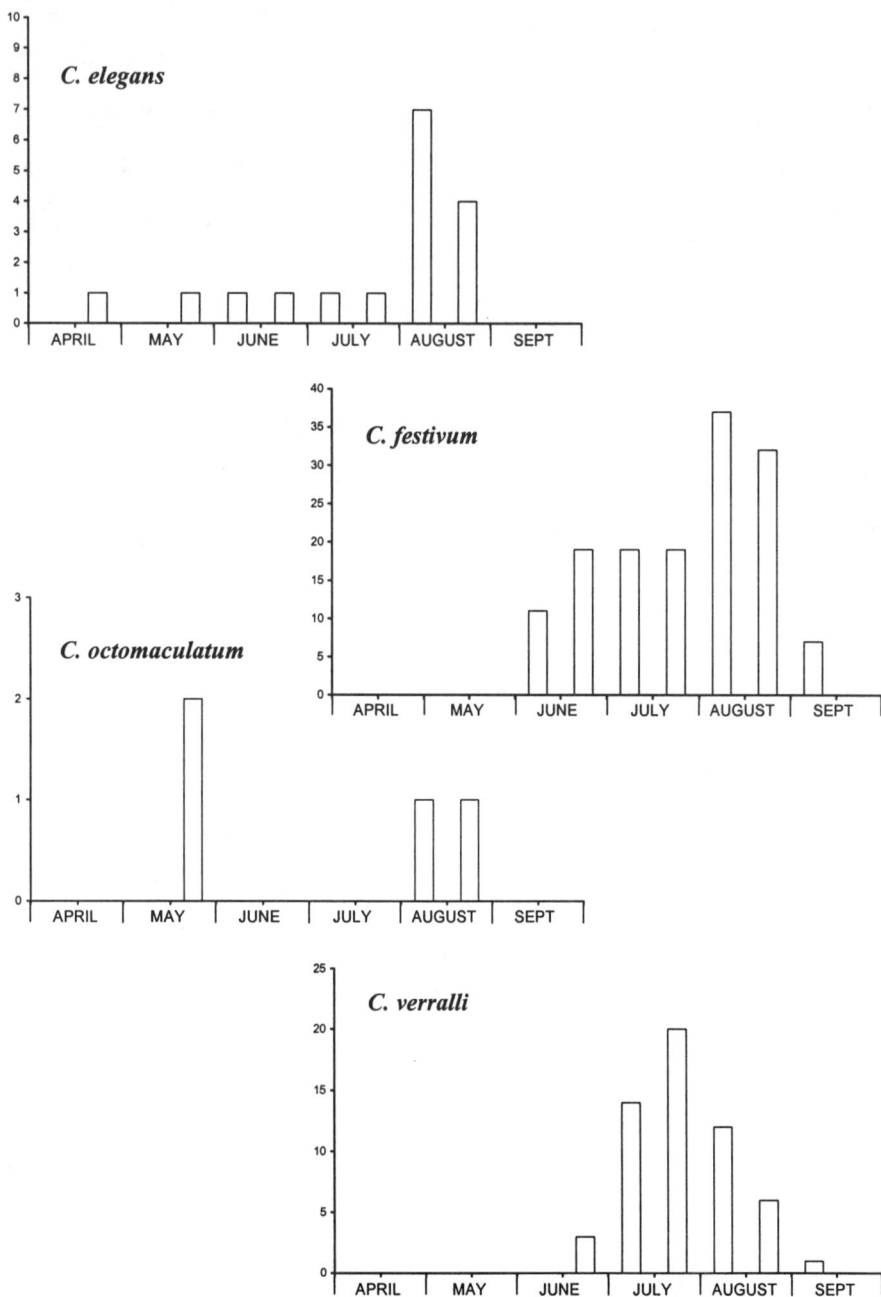

Figure 6b. **Phenology of *Chrysotoxum* species**

Genus *Dasysyrphus*

This is a group of medium-sized black-and-yellow wasp mimics with lunulate spots or bars. The larvae are generally predators on arboreal aphids, and are often associated with pines. Many adults visit flowers such as buttercups, and some can be confused in the field, especially *D. friuliensis*, *D. pinastri* and *D. venustus*.

Dasysyrphus albostriatus (Fallén, 1817)

Number of records: 259

Surrey Status: Common

Flight times: April – October

Peaks: <u>May</u>, August

This widespread and common woodland species appears to be less frequent in south-east Surrey where woodland is scarce. There is one record of a larva taken from goat willow, *Salix caprea* (Lees, 1991).

Flower visits: creeping buttercup (LP, 1950a), greater stitchwort, silverweed (LP, 1950a), cherry laurel, wood spurge (LP, 1957), field maple, cow parsley, hemlock water-dropwort, hogweed, spear thistle, common knapweed, cat's-ear, autumn hawkbit, hawkweed oxtongue, perennial sow-thistle, dandelion, hoary ragwort

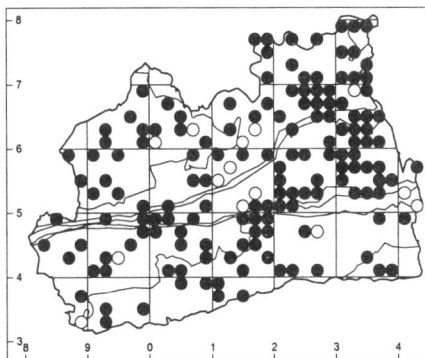

[*Dasysyrphus friuliensis* van der Goot, 1960

Number of records: 4

Surrey Status: Unconfirmed

There are a small number of records of this species, specimens of which I have not seen. Whilst I would not dismiss the possible occurrence of *D. friuliensis* in Surrey, I have doubts about these records and have not been able to examine them to verify the identifications. This species is therefore treated as unconfirmed.]

Dasysyrphus pinastri (De Geer, 1776) = *lunulatus*: auctt., misident.

Number of records: 49
Surrey Status: Local
Flight times: April – August
Peak: May

It is possible that this species is somewhat under-recorded owing to its similarity to *D. venustus*. *D. pinastri* is closely associated with conifer woodland and plantations, and is also encountered on heathland invaded by pines. It would appear to be scarce or absent from much of the Chalk. I have reservations about two records for inner London which I have not been able to examine, and these are therefore excluded from the maps.

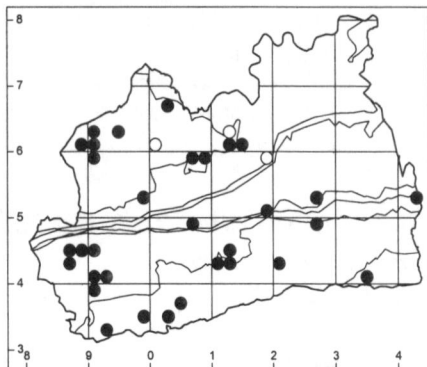

Dasysyrphus tricinctus (Fallén, 1817)   PLATE 3

Number of records: 97
Surrey Status: Local
Flight times: April – September
Peaks: May, August

This is a characteristic species of heathland and heathy woodlands in late summer when it can be quite common; elsewhere, it is more often recorded as single individuals. Its larvae are mainly associated with pine aphids and so this is a species indicative of coniferisation. Although widespread, *D. tricinctus* would seem to be virtually absent from south-east Surrey, possibly reflecting low woodland cover in keeping with other species such as *Epistrophe grossulariae* (q.v.).

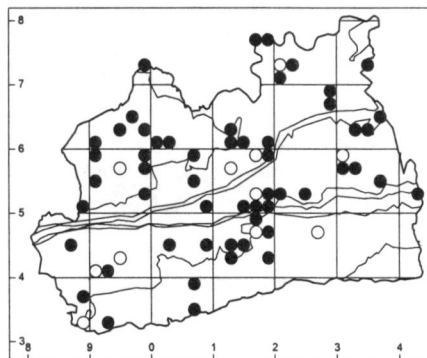

Flower visits: creeping buttercup (LP, 1950a), buttercup spp., bramble, germander speedwell (Jennings, 1895), dandelion

Dasysyrphus venustus (Meigen, 1822)   PLATE 3

Number of records: 291
Surrey Status: Common
Flight times: April – June
Peak: May

This is a widespread and common woodland hoverfly which inhabits both deciduous woodland and conifer plantations. It is particularly frequent at buttercup flowers in May and June, and would appear to be considerably more common outside the mainly urbanised London area.

Flower visits: creeping buttercup (LP, 1950a), buttercup spp., garlic mustard, rowan, firethorn, hawthorn, dandelion, bluebell (LP, 1955b)

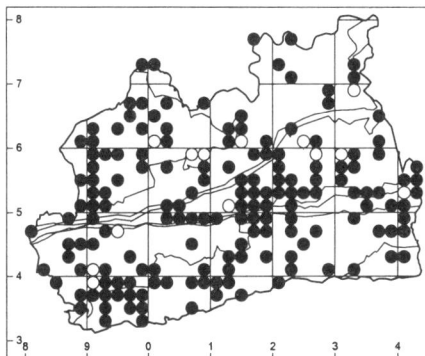

Genus *Didea*

These robust, medium to large wasp mimics often appear to be aggressive in the field. The larvae are associated with arboreal aphids, especially those on conifers.

Didea fasciata Macquart, 1834   PLATE 3

Nationally Scarce
Number of records: 61
Surrey Status: Local
Flight times: April – October
Peaks: May, August

In Surrey this is a widespread hoverfly which is frequently found at woodland and scrub edge. *D. fasciata* is widespread in the London area and considerably more common than the records in Plant (1986) suggest, but would appear to be much less common on the Chalk. There seems to be a link with heathland where the larvae are probably predators of aphids on Scots pine, *Pinus sylvestris*, but *D. fasciata* is also found around oak scrub that has invaded grassland. There is one record of a larva reared from goat willow, *Salix caprea*, at Kew Gardens, with the suggestion that the larvae were feeding on the aphid *Tuberolachnus salignus* (Lees, 1990), but in Lees (1991) this was revised to report the larva from *Salix x smithiana*. Lees (1990) also reports this species flying around an isolated willow covered in aphids on Mitcham Common. This additional

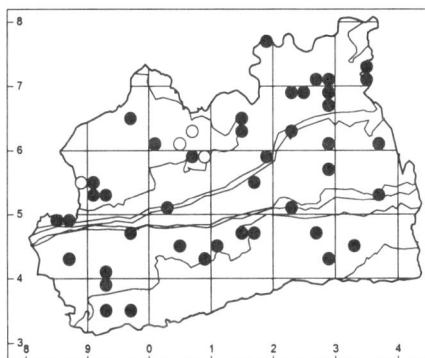

prey association may help to explain why *D. fasciata* is so widely distributed. Males are aggressive and defend territories (Morris, 1991). There is also a record of this species taken at a mercury vapour light trap (DT).

Flower visits: elder

RECORDS: **Farnham** SU8449 (5.1993, DT); **Aldershot** SU8649 (30.5.1994, RKAM); **Tilford Reeds** SU8643 (9.5.1987, 28.5.1988, GAC); **Mytchett Lake** SU8954 (1975, 1976, AES); near **Deepcut** SU9055 (10.5.1987, GAC); **Ash Ranges (Fox Hills)** SU9152 (28.8.1982, SRM); **Witley Common** SU9239 (16.5.1992), SU9340 (3.8.1995, RKAM); **Frillinghurst Wood** SU9334 (1991, JSD); **Botany Bay** SU9734 (25.5.1987, GAC); **Compton Common** SU9646 (21.5.1988, RKAM); **Chobham Longcross** SU9765 (10.5.1987, RKAM); **Horsell Common** TQ0160 (1982, AJH); **Byfleet** (22.5.1938, LP); **Weybridge** (5.7.1909, CWW); **Merrow Common** TQ0251 (26.5.1986, RKAM); above **Woodhill Sandpit** TQ0444 (29.8.1992, RKAM); **Wisley RHS Gardens** TQ0658 (21.9.1981, 23.9.1981, AJH); **Wisley & Ockham Commons** TQ0858 (1967 - 1982, AES); **Hurtwood** TQ0843 (4.9.1988, RKAM), TQ1044 (6.5.1995, RKAM); **Friday Street** TQ1245 (6.7.1985, RKAM); **Abinger Forest** TQ1446 (21.8.1993, RKAM); **Thames Ditton and Esher Golf Course** TQ1465 (28.8.1992, RKAM); **Great Oaks** TQ1562 (18.9.1988, GAC); **Leatherhead** TQ1655 (12.5.1996, RKAM); **Holmwood** (5.6.1894, FBJ); **Ashtead Common** TQ1859 (9.8.1986, GAC); **Kew Gardens** TQ1876 (2.9.1986, 10.9.1986, 2.8.1988, 26.9.1989, RBH), TQ1876 (3.6.1990, DCL); **Reigate Heath** TQ2350 (8.10.1990, AJDM); **Nonsuch Park** TQ2363 (5.5.1989, RKAM); **Cannon Hill Common** TQ2368 (1.5.1990, RKAM); **Merton Park** TQ2569 (13.8.1995, RKAM); **Wandle Valley Sewage Farm** TQ2671 (8.5.1996, 13.8.1995, RKAM); **Benting Wood** TQ2747 (3.5.1993, RKAM); **Little Woodcote** TQ2861 (6.5.1995, RKAM); **Mitcham** TQ2868 (4.5.1997, RKAM); **Mitcham Common** TQ2868 (19.10.1959, SFI), (9.5.1984, RDD/CMJ), (11.5.1989, 21.4.1990, 7.7.1990, 15.9.1990, RKAM), TQ2867 (25.4.1990, RKAM); **Horley** TQ2942 (8.8.1981, RDH); **Hooley Downs** TQ2956 (23.7.1995, RKAM); **Tooting Bec Common** TQ2971 (21.5.1995, RKAM); **Outwood** TQ3245 (10.8.1986, GAC); **Crystal Palace** TQ3471 (8.5.1995, RKAM); **Sydenham Hill Woods** TQ3472 (10.8.1988, AG); near **Godstone** TQ3652 (1993, CWP); **Selsdon Wood** TQ3661 (28.5.1990, GAC).

Didea intermedia Loew, 1854

Nationally Scarce

Number of records: 6

Surrey Status: Rare

Flight times: May – August

Peak: June

All records of *D. intermedia* are from coniferised heathland and heathland with pine scrub, reflecting its larval association with pine aphids. Although *D. intermedia* is best separated from *D. fasciata* on the colour of the halteres, a useful character which draws attention to this species in the field is that the underside appears very pale or whitish; however, the characters in Stubbs & Falk (1983) should always be used to confirm identification.

RECORDS: **Tilford Reeds** SU8643 (28.6.1987, 28.5.1988, RKAM); **Thursley** SU9040 (26.6.1989, JSD); **Puttenham Common** SU9045 (6.8.1988, RKAM); **Wisley RHS Gardens** TQ0658 (1980 - 1981, AJH); **Wisley and Ockham Commons** TQ0858 (3.8.1965, AES).

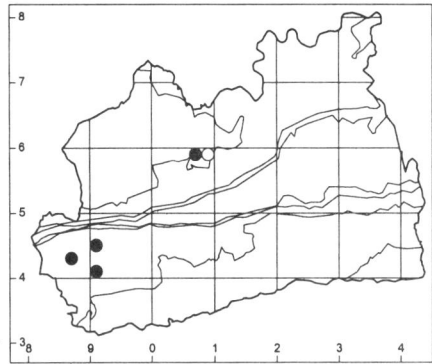

Genus *Doros*

Doros profuges (Harris, 1780) = *conopseus* (Fabricius, 1775)   PLATE 8

Red Data Book 2

Number of records: 10

Surrey Status: Rare

Flight times: June

The majority of historical records come from the Dorking Gap region of Surrey (e.g. LP, 1950b, 1952b) and suggest that *D. profuges* is a chalk downland species, but the recent record from Epsom Common indicates that its habitat preferences extend at least to the clays that overlie chalk. A specimen from Park Downs, Banstead, was observed "ovipositing on the trunk of an isolated ash tree surrounded by chalk downland and scrub" (Hawkins, 1985); this suggests that there may be an association with scrub edge. Although this fly was seen with its abdomen curved downwards to touch the trunk and, after capture, laid many eggs in a plastic tube, this posture may reflect the natural curve of the species' body, so oviposition is not proven in this instance (R. Hawkins, *pers. comm.*). *D. profuges* is probably associated with aphids in ant nests (G. Rotheray, *pers. comm.*). It is listed on the long list of the UK Biodiversity Action Plan (DoE, 1995) and is a species whose biology needs further investigation.

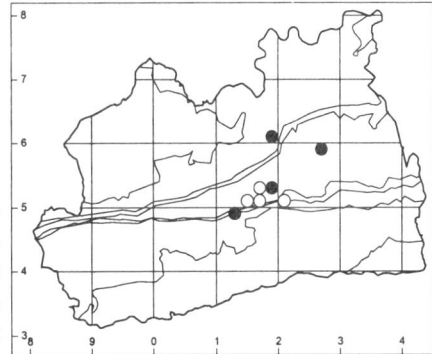

Conservation: Scrub edge may be an important habitat for this species and therefore the management of downland, where *D. profuges* is known to occur, should aim to maintain a long and varied scrub edge. Surrey is one of the strongholds for this species and if a strong population is discovered, autecological studies should be encouraged in order that its requirements are better understood.

RECORDS: **Westcott Downs** TQ1249 (9.6.1985, AJH); **Ranmore** TQ1450 (17.6.1967, SFI); **Mickleham Downs** TQ1753 (10.6.1950, 17.6.1951, PWC); **Box Hill** TQ1752 (1979, SBRC), TQ1751 (23.6.1977, SBRC); **Headley Warren** TQ1853 (12.6.1994, GAC); **Epsom Common** TQ1860 (25.6.1993, DE); **Betchworth** (2.6.1950, SW); **Banstead Park Downs** TQ2658 (16.6.1985, RDH) (17.6.1994 - visual record, RDH).

Genus *Epistrophe*

A group of medium-sized species with yellow bars and in one instance yellow spots. They may be mistaken for the commoner *Syrphus* species, but are generally narrower bodied and have a distinct orange hue. The larvae are aphid predators associated with arboreal aphids and aphids on umbellifers as far as we know. One species, *E. melanostoma*, has been newly recognised in Britain and there is another closely related species, *E. ochrostoma*, which may be found in Surrey but is otherwise known from a single British record. There are possibilities of other similar species arriving from the continent, so retention of voucher specimens is desirable.

Epistrophe diaphana (Zetterstedt, 1843)

Nationally Scarce
Number of records: 76
Surrey Status: Local
Flight times: May – August
Peak: June/July

In Verrall's time, this species was known from just one specimen from a site on the south coast and even today it is regarded as nationally scarce; however, *E. diaphana* is widespread and not uncommon in Surrey, and appears to be much better represented here than elsewhere in the country. It is a frequent visitor to umbel flowers, especially hogweed (13 out of 16 flower visit records) in partially scrubby locations and along woodland rides, but can also be found in ruderal situations throughout much of the county. The high frequency of records of visits to hogweed may

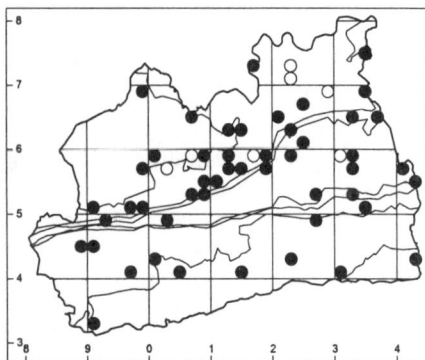

reflect an association with aphids on hogweed; it has been observed ovipositing on hogweed in Middlesex (Dobson, 1997). *E. diaphana* is apparently absent from much of the heathland in west Surrey and from the Chalk, with most records from the London clay and Low Weald.

Flower visits: cow parsley, burnet-saxifrage, hogweed, upright hedge-parsley

RECORDS: **Thundry Meadows** SU8944 (28.6.1987, GAC); **Puttenham Common** SU9045 (6.8.1988, GAC); **Normandy** SU9050 (13.8.1994, RKAM); **Almshouse Common** SU9132 (3.8.1995, RKAM); **Hog's Back** SU9248 (28.6.1987, RKAM); **Enton Green** SU9640 (30.7.1994, RKAM); **Broadstreet Common** SU9751 (11.7.1987, GAC); **Guildford, River Wey** SU9951 (3.8.1995, RKAM); **Hoe Stream** SU9956 (24.7.1994, RDH); **Gallow Hill** SU9969 (26.5.1995, RKAM); **Bramley** TQ0143 (18.6.1994, RKAM); **Maybury** TQ0159 (5.7.1987, GAC); **Merrow Downs** TQ0349 (29.8.1988, GAC); **Wey Navigation Canal** TQ0357 (11.6.1970, COH); **Smithwood Common** TQ0441 (22.7.1995, RKAM), TQ0541 (5.7.1988, GAC); **Hatchlands** TQ0652 (13.6.1986, RKAM); **Chertsey Meads** TQ0665 (23.8.1987, RKAM); **The Sheepleas** TQ0852 (6.6.1993, RKAM); near **Horsley** TQ0855 (11.6.1989, GAC); **Wisley Common** (6.1964, COH), TQ0658 (1967 - 1982, AES); **Ockham Common** TQ0858 (11.6.1989, RKAM); **Effingham** (26.8.1934, LP), TQ1055 (11.6.1994, RKAM); **Great Bookham Common** (13.7.1947, LP), (1937, COH), TQ1256 (17.6.1987, RKAM/GAC); near **Cobham** TQ1258 (9.8.1997, RDH); **Stoke D'Abernon** TQ1358 (22.6.1996, RKAM); **Jayes Park** TQ1441 (5.6.1993, RKAM); **West End Common, Esher** TQ1263 (29.6.1988, GAC); **Fetcham** TQ1457 (22.6.1996, RKAM); **Great Oaks** TQ1562 (9.8.1987, 26.6.1988, RKAM); **Eel Pie Island** TQ1672 (25.6.1989, RKAM); **Ashtead Common** (3.7.1932, LP), TQ1859 (22.6.1986, 2.7.1986, GAC); **Stane Street** TQ1956 (10.7.1989, RKAM); **Ewell** TQ2064 (10.6.1995, RKAM); **Epsom Downs** TQ2259 (22.6.1988, GAC); **Nonsuch Park** TQ2263 (11.6.1994, RDH); **Wimbledon Common** (20.7.1954, RWJU); **Edolphs Copse** TQ2343 (4.7.1987, RKAM/GAC); **Morden Park** TQ2467 (8.7.1989, RKAM); **Banstead** (8.7.1951, DJC); **Banstead Downs** TQ2560 (24.6.1989, RKAM), TQ2561 (18.6.1988, RKAM); **Earlswood Lakes** TQ2748 (5.7.1988, GAC); **Upper Gatton** TQ2753 (18.6.1988, RKAM); **Mitcham Common** TQ2868 (1971 - 1974, AES); **Fernhill** TQ3041 (4.7.1994, RDH); **Coulsdon Common** TQ3256 (7.8.1996, RDH); **Coulsdon** (19.6.1948, LP), (26.6.1948, 27.6.1948, COH); **Brewer Street** TQ3352 (19.7.1986, GAC); **Kenley Common** TQ3358 (27.6.1995, 30.6.1995, RDH); **Riddlesdown Quarry** TQ3359 (19.8.1987, GAC), **Riddlesdown** TQ3359 (9.8.1995, RDH); **Lloyd Park** TQ3364 (21.6.1996, RKAM); **Bay Pond** TQ3551 (6.7.1985, RKAM/GAC), (26.6.1988, GAC);**South Norwood Country Park** TQ3568 (2.7.1988, RKAM/GAC); **Nunhead Cemetery** TQ3575 (9.7.1994, RDH); **Threehalfpenny Wood** TQ3764 (21.6.1996, RKAM); **Tatsfield** TQ4156 (12.7.1986, GAC); near **Dormansland** TQ4243 (20.8.1994, RKAM); **Titsey Wood** TQ4254 (6.7.1986, GAC).

Epistrophe eligans (Harris, 1780)   PLATE 2

Number of records: 552
Surrey Status: Ubiquitous
Flight times: March – August
Peak: May

This common hoverfly of woodland edges is widely distributed throughout Surrey. Adults are particularly frequent around elder leaves. The relatively low density of records is because this species' flight period is quite short, making widespread recording difficult. There is some evidence that the emergence period has advanced well into April with the recent run of exceptionally warm springs, and in 1997 several individuals were recorded in late March.

Flower visits: buttercup spp., blackthorn, cherry laurel, rowan, firethorn, hawthorn, dog's mercury, wood spurge (LP, 1957), Norway maple, field maple, sycamore, cow parsley

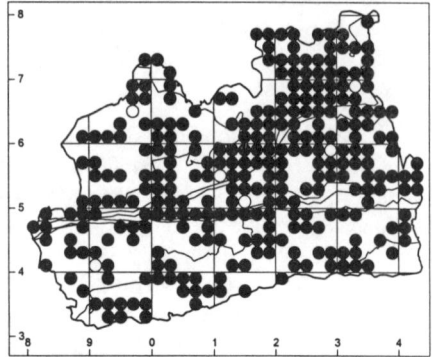

Epistrophe grossulariae (Meigen, 1822)   PLATE 1

Number of records: 240
Surrey Status: Common
Flight times: May – October
Peak: August

This is a common species which is most abundant during the hogweed season, but it can also be common at devil's-bit scabious and knapweed. It is a woodland species whose larvae seem to be associated with arboreal aphids, especially those on sycamore (G. Rotheray, *pers. comm.*), which may explain the distinct dearth of records for south-east Surrey where woodland cover is limited; this is a feature common to a number of other species which may be closely associated with woodland.

Flower visits: bramble, ivy, ground-elder, wild parsnip, hogweed, field bindweed, elder, field scabious, devil's-bit scabious, small scabious, greater knapweed, common knapweed

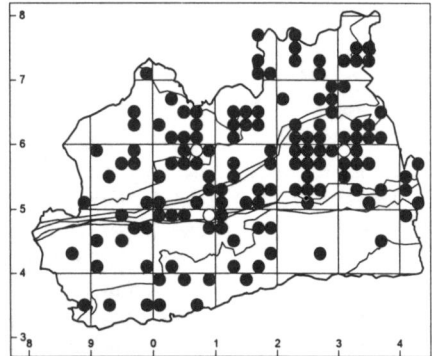

Epistrophe melanostoma (Zetterstedt, 1843)

Number of records: 19
Surrey Status: Local
Flight times: April – June
Peak: May

This species was first recognised as British by Beuk (1990) on the basis of a specimen taken at Mitcham Common (TQ2868) on 7.5.1989, but examination of material in past collections indicates that the first specimen was taken by Graham Collins at Merrow Common in 1986. In the field it looks distinctly more orange than *E. nitidicollis*.

Careful examination of *E. nitidicollis* is essential as there is considerable variation in the proportion of black scutellar hairs, with some specimens possessing just a few (see *E. nitidicollis*), making misidentification a possibility. Given this variation, the use of the scutellar hair colour may be insufficient to ensure correct identification. Care should also be taken to ensure that *E. ochrostoma* is not overlooked. On the basis of current data, it would seem that this species emerges slightly earlier than *E. nitidicollis* and is a woodland species, with most records from woodland edge or woodland rides.

Flower visits: buttercup spp., dog rose, cow parsley

RECORDS: **Frimley Playing Fields** SU8956 (1.5.1993, RKAM); **Wanborough Wood** SU9149 (29.4.1990, RKAM); **West End Common** SU9359 (29.5.1994, RKAM); **Shalford** SU9947 (7.5.1990, RKAM); **Selhurst Common** TQ0140 (25.5.1996, RKAM); **Cartbridge** TQ0156 (8.5.1993, RKAM), TQ0256 (8.5.1993, RKAM); **Merrow Common** TQ0251 (26.5.1986, GAC), (28.4.1990, RKAM); near **Clandon** TQ0450 (1.5.1993, RKAM); **Great Bookham Common** TQ1256 (9.5.1997, GAC); near **Leatherhead** TQ1756 (1.5.1993, RKAM); **East Sheen Common** TQ1974 (30.5.1992, RKAM); **Ewell** TQ2064 (10.6.1995, RKAM); **Banstead Downs** TQ2560 (15.5.1988, RKAM); **Hookwood** TQ2742 (31.5.1994, RDH); **Mitcham Common** TQ2868 (7.5.1989, PLTB); **South Croydon** TQ3363 (12.5.1996, 15.6.1996, GAC)

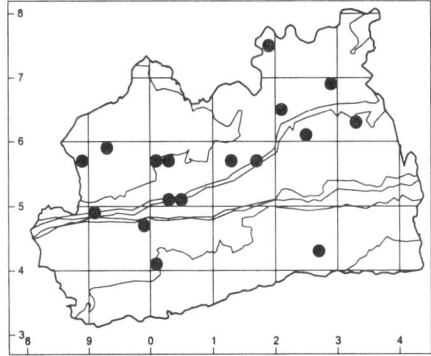

Epistrophe nitidicollis (Meigen, 1822)

Number of records: 47

Surrey Status: Local

Flight times: April – July

Peak: May

This widespread woodland hoverfly can be found sunning itself on the leaves of sycamore and other trees. This species may have become more common in the last five years, but this may also be a reflection of discovering its habits. Stubbs (1996) uses the colour of the scutellar hairs to differentiate this species from *E. melanostoma*, but considerable care needs to be exercised as scutellar hair colour is extremely variable. In my collection the breakdown of the proportion of black hairs on the scutellum of *E. nitidicollis* is as follows:

All black – 1, 75% black – 15, 50% black – 8, 25% black – 3, <10% black – 1.

In most cases, any black hairs will be towards the scutellar margin and on the disc of the scutellum.

Flower visits: bramble, hawthorn, cow parsley

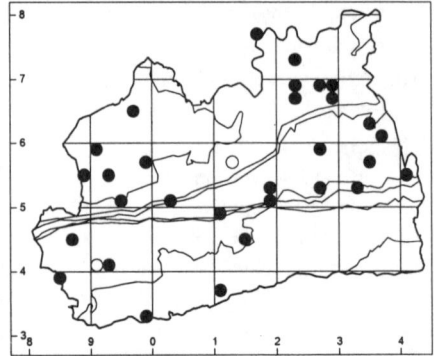

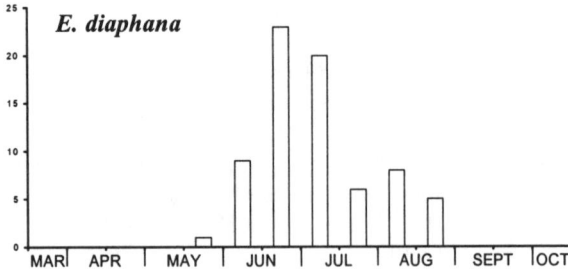

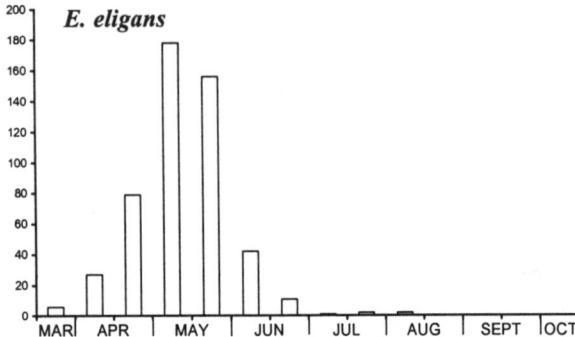

Figure 7a. **Phenology of *Epistrophe* species**

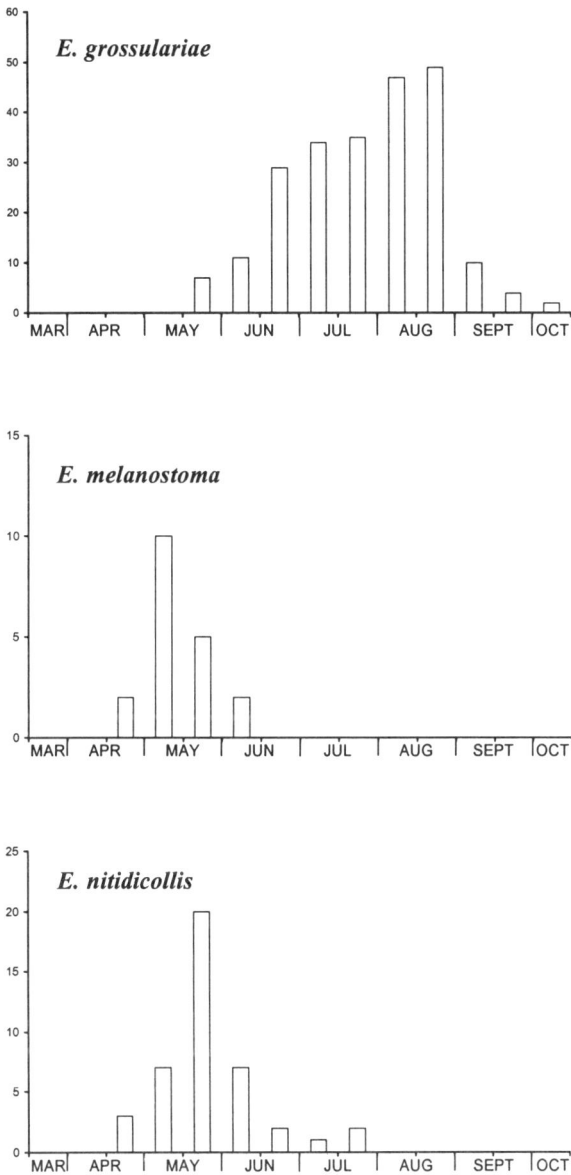

Figure 7b. Phenology of *Epistrophe* species

Genus *Episyrphus*

Episyrphus balteatus (De Geer, 1776)

Number of records: 1500
Surrey Status: Ubiquitous
Flight times: March – December
Peak: August

This is the commonest hoverfly in Surrey and one whose numbers are often bolstered by migration in late summer. Adults exhibit a considerable range of abdominal colouration from nearly pure orange to almost black, leading to confusion amongst some novice hoverfly recorders. I have one record of a female ovipositing on hogweed (RKAM) and a second of a female probably ovipositing on hogweed (RDH).

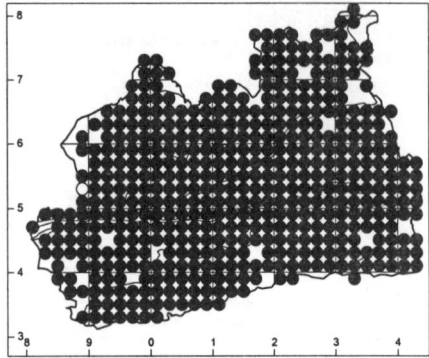

Flower visits: meadow buttercup, creeping buttercup, buttercup spp., lesser spearwort, opium poppy, common poppy, goosefoot spp., greater stitchwort, lesser stitchwort, water chickweed, redshank, knotgrass, Japanese knotweed, perforate St John's-wort, white bryony, hedge mustard, garlic mustard, winter-cress, perennial wall-rocket, charlock, weld, heather, *Erica* spp., yellow loosestrife, meadowsweet, bramble, silverweed, hawthorn, herb robert, ivy, cow parsley, burnet-saxifrage, ground-elder, fool's water-cress, wild angelica, wild parsnip, hogweed, upright hedge-parsley, wild carrot, field bindweed, hedge bindweed, large bindweed, wild marjoram, water mint, butterfly-bush, garden privet, *Hebe* sp., red bartsia, lady's bedstraw, elder, devil's-bit scabious, small scabious, welted thistle, spear thistle, marsh thistle, creeping thistle, common knapweed, nipplewort, cat's-ear, autumn hawkbit, bristly oxtongue, hawkweed oxtongue, perennial sow-thistle, smooth sow-thistle, prickly sow-thistle (LP, 1950a), dandelion, smooth hawk's-beard, common fleabane, Michaelmas-daisy, tansy, yarrow, scentless mayweed, common ragwort, hoary ragwort, ragwort spp., water-plantain

Genus *Eriozona*

Eriozona erratica (Linnaeus, 1758) = *Megasyrphus annulipes* (Zetterstedt, 1838)
Nationally Scarce
Number of records: 1
Surrey Status: Rare
There is only a single record from Chipstead (10.5.1965) on the Parmenter record cards which may represent a vagrant or migrant as *E. erratica* appears to be more common in conifer plantations in northern and western England (Ball & Morris, *in prep.*) Given that conifer associates appear to be spreading, it would be worthwhile seeking this species in highly coniferised areas such as Hurtwood or Chiddingfold.

Genus *Eupeodes*

This is the former genus *Metasyrphus*, a group of medium-sized yellow-and-black wasp mimics with bands or spots. Some are difficult to identify whilst others, such as *E. luniger,* show significant size variation. Vouchers of the more difficult species should be retained, together with any which are problematic. It is possible that further species will be recognised within this group in due course. The larvae are aphid predators.

Eupeodes corollae (Fabricius, 1794)
Number of records: 299
Surrey Status: Common
Flight times: April – September
Peak: August
This is a common hoverfly with no obvious distribution pattern. It is known to be migratory (Stubbs & Falk, 1983) and this may help to explain the widespread distribution of records and the peak occurrence in August when it is often abundant on umbel flowers.

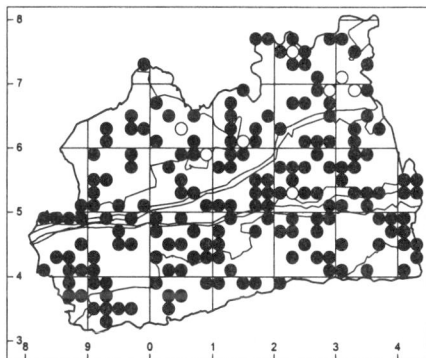

Flower visits: buttercup spp., perforate St John's-wort, charlock, heather, wood spurge (LP, 1957), burnet-saxifrage, fool's water-cress, wild parsnip, hogweed, wild carrot, field bindweed, *Hebe* sp., creeping thistle, cat's-ear (LP, 1956b), beaked hawk's-beard, common fleabane, scentless mayweed

Eupeodes latifasciatus (Macquart, 1829)

Number of records: 106
Surrey Status: Local
Flight times: April – October
Peaks: May, <u>August</u>

There are no obvious patterns to the distribution of this widespread species which is probably commoner than the records suggest. It is possibly associated with rough grassland in woodland rides, but more detailed observations are needed to confirm this. There is also a distinct dearth of records in July between the low peak in May and the high peak in August.

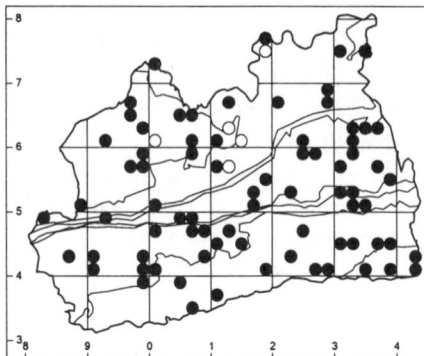

Flower visits: buttercup spp., lesser celandine, redshank, buckwheat, heather, hogweed, cat's-ear, common fleabane, Canadian goldenrod

Eupeodes latilunulatus (Collin, 1931)

Nationally Scarce
Number of records: 11
Surrey Status: Rare
Flight times: April – September
Peak: August

This is a scarce and very variable hoverfly which may prove to be part of a species complex. Many records, but not all, are from conifer plantations. Amongst material in my collection are specimens from Tugley Wood and near Capel which have bare patches on the alula, and similar variation exists amongst specimens held by Graham Collins. A record from Mitcham Common (Dunn & Johnson, 1984) is now known to be incorrect.

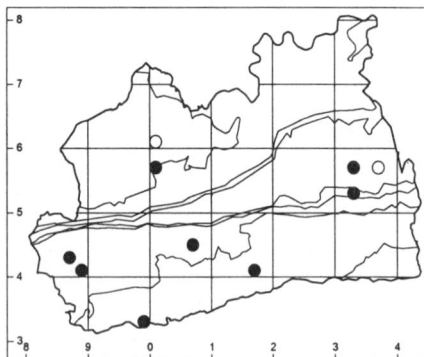

RECORDS: **Tilford Reeds** SU8643 (13.6.1987, GAC), (28.6.1987, RKAM); **Hankley Common** SU8841 (18.4.1993, JRD); **Tugley Wood** SU9833 (17.7.1993, RKAM); **Horsell Common** (13.5.1956, LP); **White Rose Lane NR** TQ0157 (2.9.1995, AJH); **Hurtwood** TQ0644 (22.7.1995, RKAM); near **Capel** TQ1741 (21.8.1993, RKAM); **Coulsdon Common** TQ3257 (5.8.1993, RKAM); **Worms Heath** (9.5.1942, LP); **Brewer Street** TQ3352 (22.8.1987, GAC).

Eupeodes luniger (Meigen, 1822)

Number of records: 471

Surrey Status: Ubiquitous

Flight times: March – November

Peaks: May, <u>August</u>

This common migratory hoverfly can vary greatly in size and there may be other differences which could lead to species splits in due course. This can be a common spring species which may help to explain the concentration of records in the London area. Although thought to be partially migratory (Stubbs & Falk, 1983), *E. luniger* is certainly resident as shown by the report by Champion (1912) of 3 adults reared from larvae taken in May 1912 from young pines on Horsell Common; a record of larvae swept from scentless mayweed at Crowhurst Lane End (TQ3848) on 12.7.1986 (GAC); and of a female laying eggs on bramble in Lambeth (TQ3178) on 10.4.1997 (RDH).

Flower visits: creeping buttercup (LP, 1950a), buttercup spp., chickweed, lesser stitchwort, redshank, knotgrass, wood dock, perforate St John's-wort, hedge mustard, garlic mustard, charlock, heather, blackthorn (LP, 1950a), cherry laurel, firethorn, hawthorn (LP, 1957), red clover, wood spurge (LP, 1957), ragged robin (LP, 1950a), cow parsley, wild parsnip (LP, 1950a), pepper-saxifrage (LP, 1950a), wild angelica, hogweed, wild carrot, field bindweed, gipsywort, ribwort plantain, butterfly-bush, *Hebe* sp., round-headed rampion, common knapweed, cat's-ear, autumn hawkbit, hawkweed oxtongue, smooth hawk's-beard, oxeye daisy, hoary ragwort (LP, 1950a), bluebell (LP, 1955b)

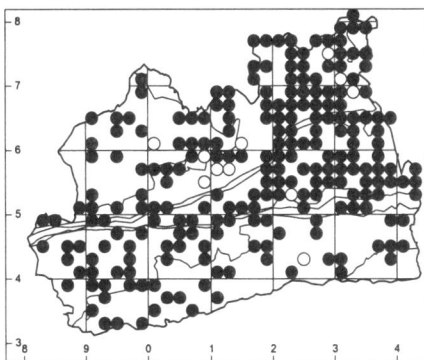

Eupeodes nitens (Zetterstedt, 1843)

Nationally Scarce

Number of records: 6

Surrey Status: Rare

Flight times: April – June

Peak: May

This scarce hoverfly has been found both in deciduous woodlands and in conifer plantations. It is possibly overlooked as it is very similar to other common *Eupeodes* species, although the males appear to be distinctly hairy.

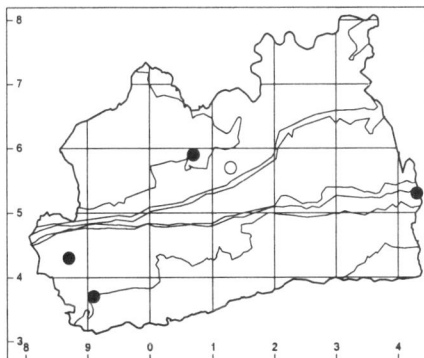

RECORDS: **Tilford Reeds** SU8643 (28.6.1987, RKAM/GAC); **Park Copse** SU9137 (9.5.1987, GAC); **Wisley Common** TQ0658 (26.4.1986, GAC); **Great Bookham Common** (LP); **Limpsfield Chart** TQ4252 (6.5.1987, GAC).

Genus *Leucozona*

A group of three medium to large species, two of which have blue markings whilst the third has distinctive pale whitish yellow markings. The larvae are aphid predators.

Leucozona glaucia (Linnaeus, 1758)   PLATE 1

Number of records: 114

Surrey Status: Local

Flight times: June – September

Peak: August

This hoverfly seems to be closely associated with hogweed in woodland rides, and its larvae are reported to be predators of ground-layer aphids (Rotheray, 1993). There is a clear concentration of records along the North Downs and a lack of records from urbanised London, which is probably due to a shortage of suitable woodland and not the level of recording effort.

Flower visits: wild angelica, wild parsnip, hogweed

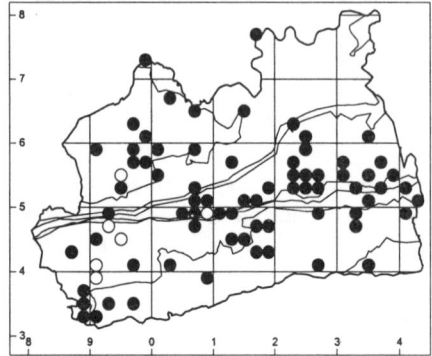

Leucozona laternaria (Müller, 1776)

Number of records: 143

Surrey Status: Local

Flight times: June – August

Peak: July

Another of the hoverflies which frequent hogweed in wooded localities, *L. laternaria* is more widely distributed than *L. glaucia*, suggesting more catholic habitat preferences. Nonetheless, it too exhibits a preference for wooded areas in central Surrey and an apparent absence from much of the London suburbs.

Flower visits: hogweed

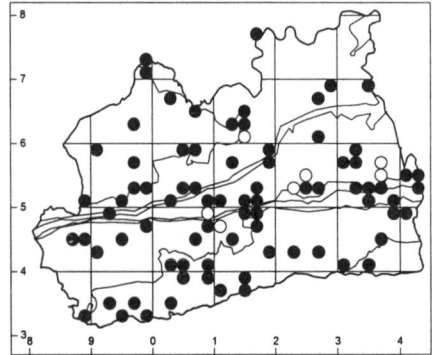

Leucozona lucorum (Linnaeus, 1758)   PLATE 2

Number of records: 441

Surrey Status: Ubiquitous

Flight times: April – August

Peak: May

This is one of the commoner spring hoverflies which mainly frequents hedgerows and woodland rides. Although undoubtedly fairly ubiquitous in the wider countryside, it is clearly much scarcer in the London area where suitable semi-natural habitat is scarce.

Flower visits: creeping buttercup, bulbous buttercup, buttercup spp., lesser celandine, greater stitchwort, garlic mustard, cuckoo-flower, rowan, hawthorn, wood spurge (LP, 1957) field maple, herb robert, cow parsley, ground-elder, hemlock water-dropwort, hogweed, white dead-nettle, dandelion

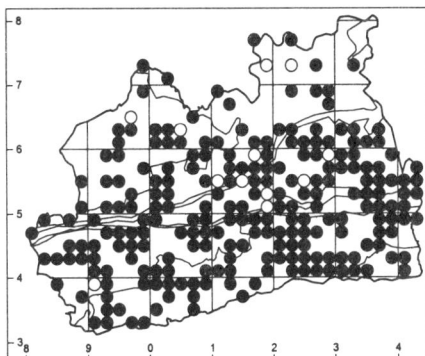

Genus *Melangyna*

These are relatively narrow-bodied medium-sized hoverflies with a range of markings from almost uniformly dark bodies to others with large whitish or yellow spots. The larvae are aphid predators which are apparently more prey-specific than other hoverflies (Rotheray, 1993). Some adults are very difficult to separate and it is essential that the more difficult species are represented by vouchers.

Melangyna barbifrons (Fallén, 1817)

Nationally Scarce

Number of records: 2

Surrey Status: Rare

Flight times: April – June

This early spring species appears to be extremely uncommon with just the single post-1985 record.

Flower visits: sallow

RECORDS: **Tilford Reeds** SU8643 (3.4.1988, GAC); **Great Bookham Common** (12.6.1948, LP).

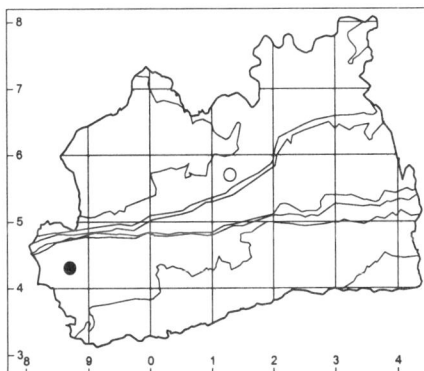

Melangyna cincta (Fallén, 1817)

Number of records: 86
Surrey Status: Local
Flight times: March – August
Peak: May

One of the best ways of recording this species is to examine the leaves and flowers of sycamore which appear to be a major lure. *M. cincta* is mainly a woodland species which occurs in very variable numbers from year to year. The larvae are reported to be principally associated with beech (Rotheray, 1993), but have also been recorded from oak and lime (G. Rotheray, *pers. comm.*).

Flower visits: lesser celandine, sycamore, dandelion

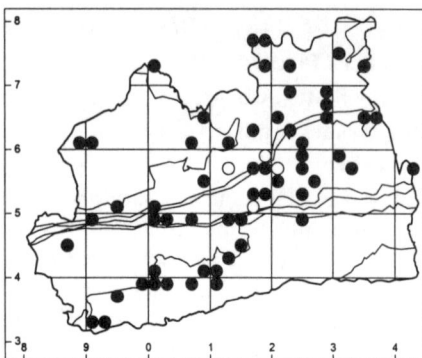

Melangyna compositarum (Verrall, 1873)

Number of records: 10
Surrey Status: Rare
Flight times: May – August
Peak: May

This species, if indeed it is separate from *M. labiatarum*, is very difficult to identify and does not appear to have consistent characters. My records are for males which appear to conform to the characters in Stubbs & Falk (1983), but even then I have reservations about them and they must be treated with caution.

Flower visits: wild parsnip, hogweed, wild carrot

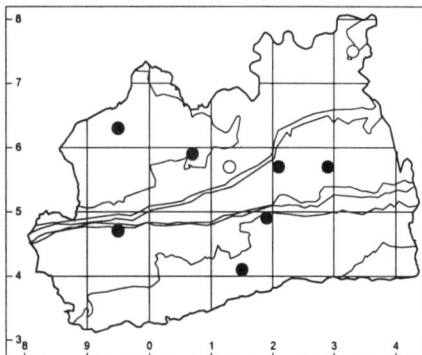

Melangyna labiatarum (Verrall, 1901)   PLATE 2

Number of records: 171
Surrey Status: Common
Flight times: May – October
Peaks: <u>June</u>, August

This common woodland-edge species is frequently seen at the flowers of hogweed and wild angelica. The males have been observed defending territories in sunny locations, hovering at about head height (RKAM). The apparent scarcity of this species in west Surrey is difficult to explain, but it is possible that it favours damper woodlands.

Flower visits: cow parsley, ground-elder, hemlock water-dropwort, wild angelica, hogweed, upright hedge-parsley, wild carrot

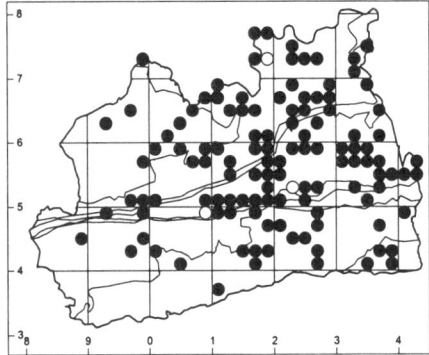

Melangyna lasiophthalma (Zetterstedt, 1843)

Number of records: 54
Surrey Status: Local
Flight times: March – May
Peak: April

For some reason, *M. lasiophthalma* appears in widely fluctuating numbers from year to year. It is principally a woodland species which can also occur in numbers on pine-infested heathland (flying around pine trees) and is often common at sallow catkins. The apparent absence of this species from south-east Surrey may be related to low woodland cover, or may be a function of recording effort at a time of year when recording can be difficult.

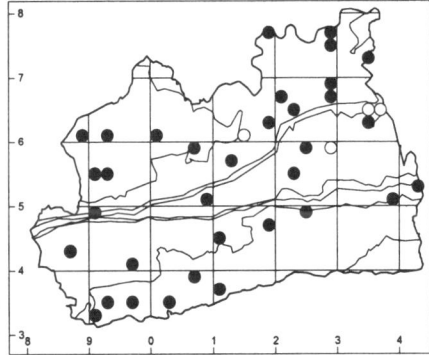

Flower visits: creeping buttercup (LP, 1950a), sallow (LP, 1950a), bird cherry, bluebell

Melangyna quadrimaculata (Verrall, 1873)

Nationally Scarce
Number of records: 11
Surrey Status: Rare
Flight times: March – May
Peak: March/April

The apparent scarcity of this spring species may be real, but it is quite likely to be related to the early flight period and poor weather in early spring, which makes recording very difficult. The attraction of this species to yellow objects is clearly illustrated by the specimen at Dawcombe which was attracted to a yellow oil can. Rotheray (1993) reports that the larvae are associated with aphids on firs, *Abies* spp., and this may also help to explain the relatively sparse distribution of records.

Flower visits: sallow, bird cherry

RECORDS: **Tilford Reeds** SU8643 (3.4.1988, GAC); **Camberley** (6.5.1922, specimen in Plumstead Museum, collector uncertain); **Lythe Hill** SU9232 (1989, JSD) **Claremont** TQ1263 (no date, JWY); **Esher and Oxshott Commons** (no date, JWY); **Wotton Woods** TQ1147 (3.4.1988, RKAM); **Box Hill** TQ1852 (13.3.1993, RKAM); **Headley Warren** TQ1853 (17.3.1998, at MVL, GAC); **Dawcombe** TQ2152 (13.3.1988, RKAM); **Sydenham Hill Woods** TQ3472 (8.4.1988, AG); **Limpsfield Common** (1.3.1945, LP).

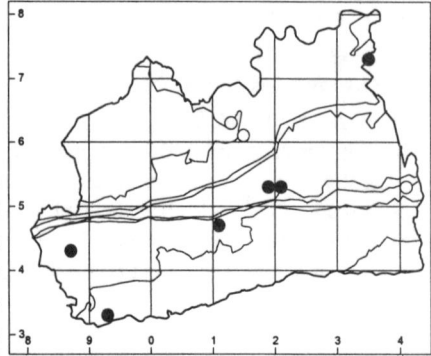

Melangyna umbellatarum (Fabricius, 1794)

Number of records: 45
Surrey Status: Local
Flight times: June – September
Peak: August

This widespread but generally scarce species is rarely seen in numbers. The larvae are associated with aphids on umbellifers (Rotheray, 1993). The adults, which are frequently attracted to umbellifers, may be associated in particular with ruderal woodland edge with hogweed.

Flower visits: hogweed, burnet-saxifrage (LP, 1966)

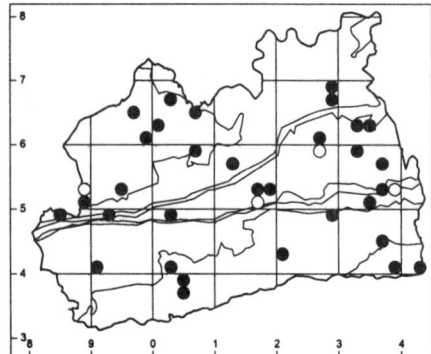

Genus *Meligramma*

This genus has recently been revised (Chandler, *in press*) with *M. cincta* moved to *Melangyna* and *Epistrophella euchroma* moved to *Meligramma*. The three British species are very different in appearance, but are relatively narrow-bodied with yellow or whitish spots. The larvae are predators of arboreal aphids.

Meligramma euchromum (Kowarz, 1885)   PLATE 15

= *Epistrophella euchroma* (Kowarz, 1885)

Red Data Book 3

Number of records: 21

Surrey Status: Rare

Flight times: April – June

Peak: May

Whilst this species has been identified by Stubbs (1982) as an ancient woodland indicator, this must be questioned as it has been reared from a puparium in a suburban garden (GAC), and other sites which have yielded this hoverfly range from ancient woodland to recent birch scrub. *M. euchromum* is most frequently met with at sunlit leaves in woodland, and may in fact be under-recorded.

Flower visits: blackthorn, hawthorn (LP, 1950a), wood spurge (LP, 1954)

RECORDS: **Olddean Common** SU8861 (7.5.1988, GAC); **Witley Common** SU9339 (1.5.1994, RKAM); **Merrow Common** TQ0251 (28.4.1990, RKAM); **Byfleet** TQ0661 (14.4.1995, GAC); **Great Bookham Common** (12.5.1946, 2.5.1949, 10.5.1953, LP); **Ashtead Common** TQ1859 (2.5.1994, NDF), (26.4.1987, RKAM/GAC); **Glovers Wood** TQ2240 (3.5.1992, RKAM/GAC); **Cannon Hill Common** TQ2368 (1.5.1990, RKAM); **Banstead Wood** TQ2658 (no date, LP); **Chipstead** (19.6.1948, LP); **Coulsdon** (no date, LP); **Riddlesdown** TQ3260 (7.5.1996, RDH); **South Croydon** TQ3363 (puparium 3.1993, adult reared, GAC); **Caterham** TQ3457 (19.5.1990, RKAM).

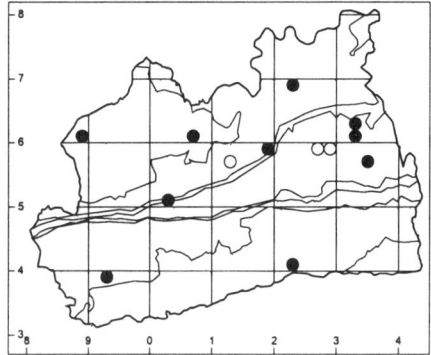

Meligramma guttatum (Fallén, 1817)

Nationally Scarce
Number of records: 5
Surrey Status: Rare
Flight times: May – July
Peak: June/July

Most of the records of this species are from west Surrey in the vicinity of the various canals. The locations where I have encountered this fly are wet and support extensive willow scrub or woodland, suggesting a possible association with willows, even though Rotheray (1993) quotes sycamore aphids as the only known prey association.

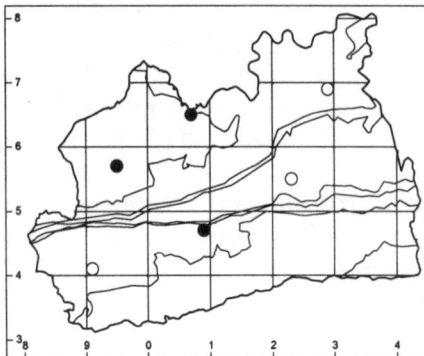

Flower visits: hogweed

RECORDS: **Thursley Common** SU9041 (28.5.1966, AES); **Basingstoke Canal** SU9557 (16.7.1994, RKAM); **Coxes Lock** TQ0664 (5.7.1987, RKAM); **Gomshall** TQ0847 (13.6.1986, RKAM); **Banstead Heath** TQ2255 (26.7.1964, AWJ); **Mitcham Common** (6.6.1949, LP).

Meligramma trianguliferum (Zetterstedt, 1843)

Nationally Scarce
Number of records: 22
Surrey Status: Scarce
Flight times: April – September
Peak: May

It is possible that this species has been overlooked in the past, leading to the status conferred on it. *M. trianguliferum* has been found in a wide variety of habitats, especially woodland edge, and is another of the species which can be found by studying sunlit leaves. The bias towards the environs of London may be true, but it is far more likely that this reflects recorder effort at a time of year when long journeys are not always worthwhile. Curiously, there does appear to be a negative correlation with habitat on chalk, from which *M. trianguliferum* is seemingly absent.

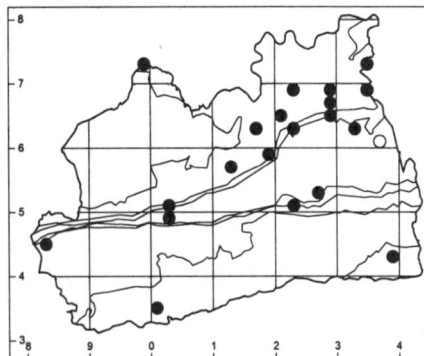

RECORDS: near **Farnham** SU8345 (23.5.1993, RKAM); **Coopers Hill** SU9972 (14.5.1988, RKAM); **Dunsfold Common** TQ0035 (7.5.1990, RKAM); **Merrow Common** TQ0251 (10.5.1989, GAC); **St Martha's Hill** TQ0348 (16.5.1987, RKAM); **Great Bookham Common** TQ1256 (10.5.1989, GAC); **Elmbridge Common** TQ1663 (12.5.1996, RKAM); **Ashtead Common** TQ1859 (26.4.1987, RKAM/GAC); **Worcester Park** TQ2065 (4.5.1987, RKAM); **Reigate Heath** TQ2350 (8.5.1989, RKAM);

Nonsuch Park TQ2363 (22.5.1990, RKAM); **Cannon Hill Common** TQ2368 (1.5.1990, RKAM); **Upper Gatton Park** TQ2652 (29.4.1994, RKAM); **Beddington Park** TQ2865 (13.5.1989, PLTB); **Mitcham Common** TQ2868 (7.5.1989, PLTB), TQ2867 (27.4.1990, RKAM); **South Croydon** TQ3363 (1.9.1985, GAC); **Sydenham Hill Woods** TQ3472 (1988, AG); **South Norwood Country Park** TQ3568 (2.7.1988, GAC); **Selsdon Wood** TQ3661 (29.5.1966, RLC); **Mill Wood, Lingfield** TQ3942 (30.5.1994, RDH).

Genus *Meliscaeva*

This is a genus of two species, both of which are characteristic, being medium-sized with long narrow bodies and wings that extend beyond the abdomen. The abdominal markings are variable and an intermediate form occurs which may cause confusion. The larvae are generally associated with arboreal and shrub aphids, but are also known to be associated with umbellifer aphids (*M. auricollis*) and psyllids (*M. cinctella*).

Meliscaeva auricollis (Meigen, 1822)

Number of records: 182
Surrey Status: Common
Flight times: January – December
Peak: June

This widespread species can often be found on sunlit leaves in the spring. The high concentration of records in the London area may reflect this, because *M. auricollis* is often common in the spring when I am prone to do short journeys to take advantage of breaks in the weather. It exhibits strong seasonal dimorphism, with a much darker spring brood. There is one record of this species as a prey item of the robber fly *Machimus atricapillus* (Diptera, Asilidae) (LP, 1968).

Flower visits: meadow buttercup, creeping buttercup, buttercup spp., lesser spearwort (LP, 1950a), lesser stitchwort, hedge mustard, garlic mustard, horse-radish, heather, blackthorn, bird cherry, cherry laurel, field maple, cow parsley, wild parsnip, hogweed, *Hebe* sp., nipplewort, cat's-ear, autumn hawkbit, common ragwort

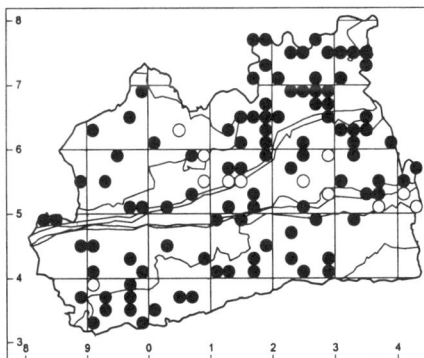

Meliscaeva cinctella (Zetterstedt, 1843)

Number of records: 357
Surrey Status: Common
Flight times: April – September
Peaks: May, <u>August</u>

This common woodland hoverfly is rarely seen in London where its low frequency is real and not a reflection of recording effort. Elsewhere, it is common at sycamore flowers and basks on the leaves. Careful examination of specimens is recommended because there is a possibility of confusion with *Melangyna cincta*.

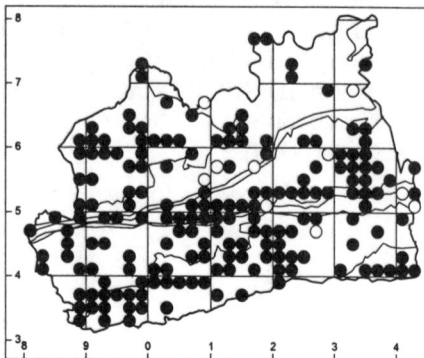

Flower visits: marsh marigold, garlic mustard, rowan, hawthorn, sycamore, ivy, wild angelica, wild parsnip, hogweed, upright hedge-parsley, elder (LP, 1950a), yarrow, ragwort spp.

Genus *Parasyrphus*

This is a group of yellow-and-black medium-sized hoverflies which may be overlooked as *Syrphus* spp. They are generally narrower bodied, however. Most are straightforward to identify, but some very similar species can occur together and it is wise to retain a few specimens to ensure that other species are not overlooked. The larvae are generally predatory on pine aphids, although *P. nigritarsis,* which is not known from Surrey, is a predator on the larvae of leaf beetles (Chrysomelidae).

Parasyrphus annulatus (Zetterstedt, 1838)

Number of records: 12
Surrey Status: Rare
Flight times: April – July
Peak: May/June

The distribution of *P. annulatus* seems to mirror the main areas of conifer plantation in central Surrey, with scattered outliers both east and west, again associated with plantations rather than heathland pines. Adults are often taken flying around the foliage of conifers, rather than at flowers.

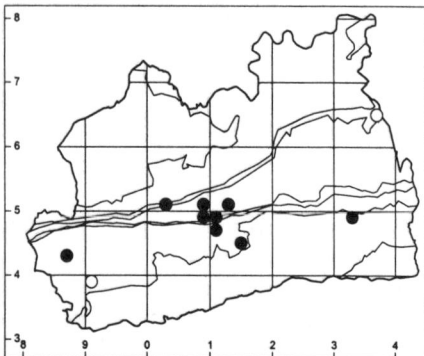

Flower visits: cherry laurel

RECORDS: **Tilford Reeds** SU8643 (9.5.1987, RKAM); **Cosford Mill** SU9139

(11.5.1969, PJC); **Merrow Common** TQ0251 (5.7.1987, GAC); **Hackhurst Downs** TQ0948 (21.6.1986, RKAM); **Effingham Forest** TQ0950 (16.6.1988, RKAM); **Wotton Woods** TQ1147 (14.6.1986, RKAM); **Oaken Grove** TQ1149 (15.6.1987, RKAM); **Ranmore Common** TQ1350 (29.4.1987, GAC); **Redlands Wood** TQ1545 (5.6.1993, RKAM); **Bransland Wood** TQ3248 (31.5.1987, RKAM/GAC); **Shirley** (5.5.1872, GHV).

Parasyrphus lineola (Zetterstedt, 1843)

Number of records: 4

Surrey Status: Rare

Flight times: April – September

There would appear to be no obvious pattern to the distribution of *P. lineola* which should be associated with conifer plantations. Neither Brewer Street nor Merrow Common support such habitats, however. It is very likely that this species, together with other *Parasyrphus*, has been overlooked.

RECORDS: **Eighty Acre Copse, Chiddingfold** SU9734 (24.4.1985, GAC); **Merrow Common** TQ0251 (5.7.1987, GAC); **Great Bookham Common** (26.6.1960, LP); **Brewer Street** TQ3352 (3.9.1988, GAC) .

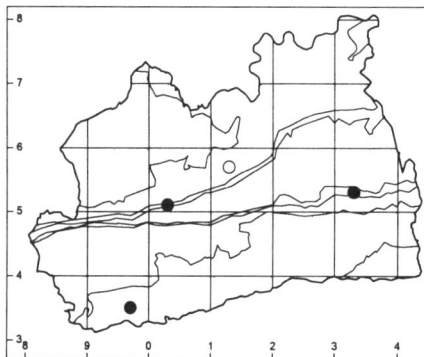

Parasyrphus malinellus (Collin, 1952)

Number of records: 4

Surrey Status: Rare

Flight times: May – June

Parasyrphus malinellus appears to be associated with coniferisation and pine invasion on heathland, and is likely to be found in conifer plantations elsewhere.

RECORDS: **Netley Heath** TQ0749 (8.6.1986, GAC); **Silent Pool** TQ0648 (16.5.1992, RKAM); **Leith Hill** TQ1343 (8.5.1989, RKAM); **Oxshott Heath** TQ1461 (1.5.1989, PLTB).

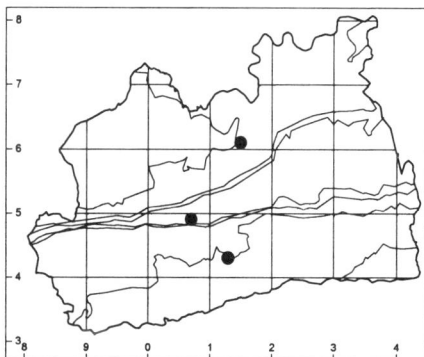

Parasyrphus punctulatus (Verrall, 1873)

Number of records: 77
Surrey Status: Local
Flight times: March – August
Peak: May

Although rarely common, this is a widely distributed hoverfly which often occurs at sallow catkins and blackthorn blossom. The heaviest concentration of records is from the Winterfold Forest area where there are extensive conifer plantations. Many other records are from heathland and the conifer plantations of the Chiddingfold Forest. There are remarkably few records from the Low Weald, possibly reflecting the lower density of conifer plantations in this area. The spread of records is somewhat wider than the distribution of conifer woodland and therefore it must be assumed that a wider range of aphid prey is utilised.

Flower visits: sallow, garlic mustard, blackthorn, bird cherry, hawthorn

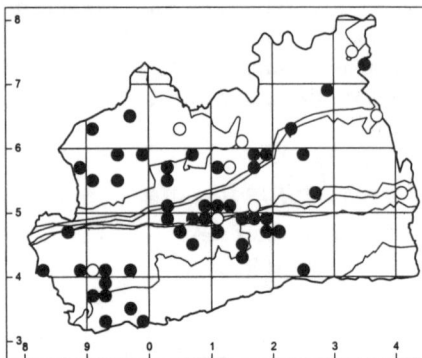

Parasyrphus vittiger (Zetterstedt, 1843)

Number of records: 23
Surrey Status: Scarce
Flight times: April – September
Peaks: May, September

When working heathland in late summer, *P. vittiger* is one of a small number of species that almost invariably occur in the sample. The distribution map suggests that it has a closer association with sandy, heathy areas, where Scots pine is the principal conifer, than other *Parasyrphus* species; thus this is not a true heathland associate, but is indicative of conifer invasion.

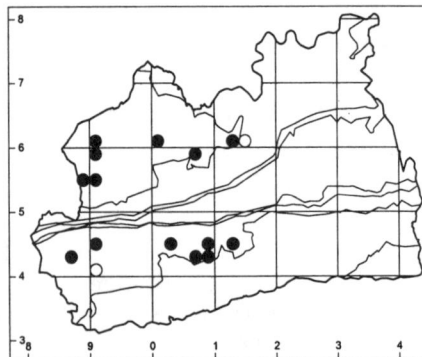

Genus *Scaeva*

These are large and characteristic species with distinct whitish bars or lunulate spots. Both are at least partially migratory, and *S. pyrastri* is commonly associated with the flux of summer migrants including *Episyrphus balteatus* and *Eupeodes* spp. The larvae are aphid predators.

Scaeva pyrastri (Linnaeus, 1758)

Number of records: 228
Surrey Status: Common
Flight times: June – September
Peak: August

This is a migratory species whose numbers fluctuate considerably from year to year. It is most frequently seen at umbel flowers, especially those of wild parsnip. It must also be partially resident as there is a record of "numerous larvae on the aphids on pine" at Oxshott (Uffen, 1958) and it has been reared from larvae swept from scentless mayweed at Crowhurst Lane End, TQ3848 (12.6.1986 (GAC). I have a melanic specimen from Alderstead Heath, TQ3055, on 11.8.1985.

Flower visits: traveller's-joy, heather, burnet-saxifrage, wild parsnip, hogweed, upright hedge-parsley, field bindweed, wild marjoram, spear thistle, common knapweed, perennial sow-thistle (LP, 1950a), common fleabane, common ragwort, ragwort spp.

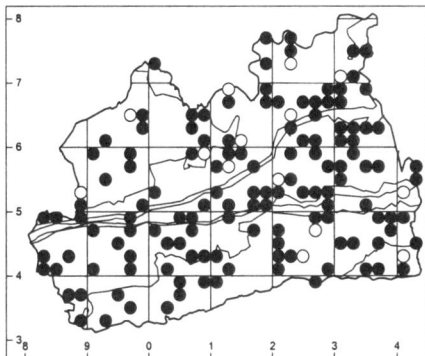

Scaeva selenitica (Meigen, 1822)

Number of records: 28
Surrey Status: Scarce
Flight times: May – September
Peaks: June, <u>August</u>

Although *S. selenitica* is thought to be partially migratory, there would appear to be quite a strong correlation with pine-infested heathland and conifer plantations, which must suggest that it is at least occasionally resident.

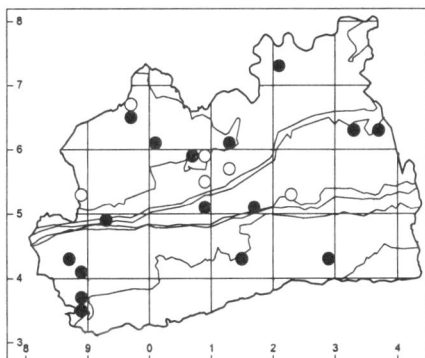

Genus *Sphaerophoria*

Only males of this genus can be identified with confidence, and pairs caught *in copula* should be retained so that future efforts to produce keys to the females can draw on a good volume of material. Since Stubbs & Falk (1983), two further species, *S. potentillae* and *S. bankowskae*, have been added to the British list. A new key in Stubbs (1996) covers all the known British species. Voucher specimens should be retained for all species except *S. scripta,* as further additions to the British list are possible (Speight, 1988). The larvae are predatory on a range of ground-layer aphids.

Sphaerophoria batava Goeldlin de Tiefenau, 1974

Number of records: 47
Surrey Status: Local
Flight times: May – September
Peaks: June, <u>August</u>

The distribution of *S. batava* is a bit of a conundrum, for it seems to be split between the heathlands of west Surrey and the gravel deposits overlying the wealden clays of south and south-west Surrey (Lousley, 1976). This is one of a number of species which is frequently found at tormentil flowers and it may be that these are the clue to its habitat requirements. Whilst *S. batava* is similar to *S. taeniata* in many respects, there are strong differences in their habitat associations; *S. batava* is found on *Calluna* heath whereas *S. taeniata* is seemingly more common on wetter clay sites. Stubbs & Falk (1983) draw attention to an additional difference between *S. batava* and *S. taeniata*: that the scutellar hairs are black in *S. batava*. They are in fact rather variable as seen from the following analysis of specimens in my collection:

All black – 15, 75% black – 12, 50% black – 4, 25% black – 5, <10% black – 1.

I have one specimen with just two black scutellar hairs. Inexperienced recorders should therefore exercise extreme caution when using scutellar hair colour as confirmation of this species' identity.

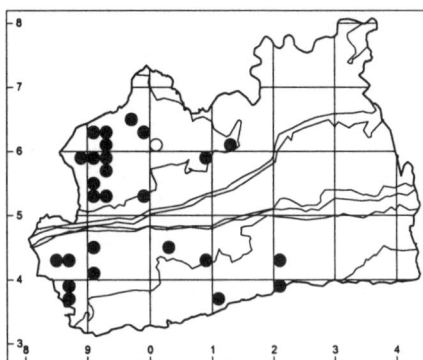

Sphaerophoria fatarum Goeldlin de Tiefenau, 1989

= *abbreviata*: auctt., misident.

Number of records: 22

Surrey Status: Scarce

Flight times: May – September

Peaks: <u>May</u>, August/September

Sphaerophoria fatarum is widely distributed on *Calluna* heath, but needs to be searched for; the best way of recording it is by sweeping the heather. It is generally wise to retain a number of specimens as this species is often found with *S. philanthus*. The records from Wimbledon Common are an addition to the list for the London area (see Plant, 1986). It has two distinct peaks, being most frequent in May, absent in June and peaking again in August.

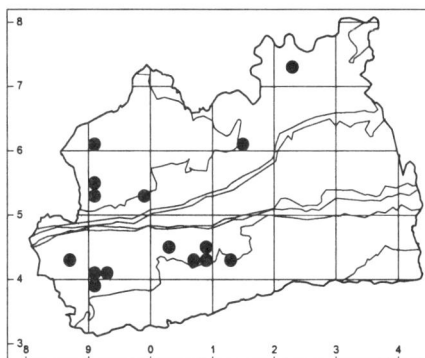

Sphaerophoria interrupta (Fabricius, 1805)

= *menthastri*: Vockeroth, 1963, misident.

Number of records: 35

Surrey Status: Local

Flight times: May – September

Peaks: May, <u>August</u>

Apart from *S. scripta* and *S. taeniata*, *S. interrupta* appears to be the most cosmopolitan of the genus and one which cannot readily be ascribed to a particular habitat. Records are scattered from chalk downland to damp clay grassland, with no obvious pattern to the distribution, except that it is seemingly absent from much of the London area and scarce on heathland. I have not accepted records of this species prior to Stubbs & Falk (1983), given that previous keys did not properly differentiate between this and related species.

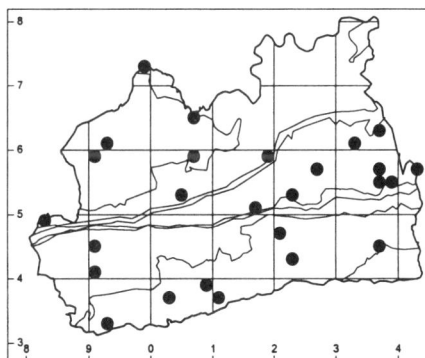

Flower visits: lady's bedstraw

Sphaerophoria philanthus (Meigen, 1822)

Number of records: 15
Surrey Status: Local
Flight times: May – September
Peak: August

Although this is principally a heathland species, there is an intriguing record from chalk downland at Box Hill (TQ1852) on 6.8.1987, which may be an individual that had strayed from nearby Headley Heath. *S. philanthus* is most frequently recorded by sweeping *Calluna* heath and is probably under-recorded, given the extent of heathland in west Surrey.

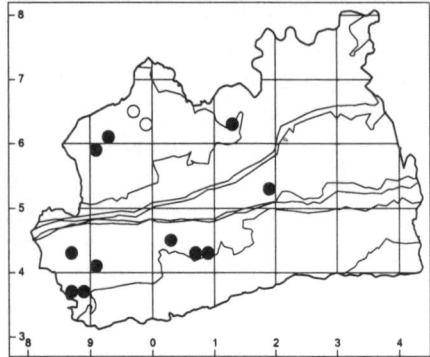

Sphaerophoria rueppellii (Wiedemann, 1830)

Number of records: 28
Surrey Status: Local
Flight times: May – September
Peak: August

Personal observations suggest that this species is as closely associated with ruderal situations as it is with prime habitat. Most records are from the London Clay and associated Thames gravels, which is consistent with this species' close association with the Thames Estuary (Ball & Morris, *in prep.*). Dobson (1992) reported this species ovipositing on great willowherb, *Epilobium hirsutum* (not in Surrey), and this behaviour has also been observed at a number of locations in Surrey, such as at Leatherhead where females, in the company of a number of males, were observed ovipositing on great willowherb which was covered with aphids (RKAM). Indeed this hoverfly has been met with as frequently in such locations as at more widely recognised lures such as redshank.

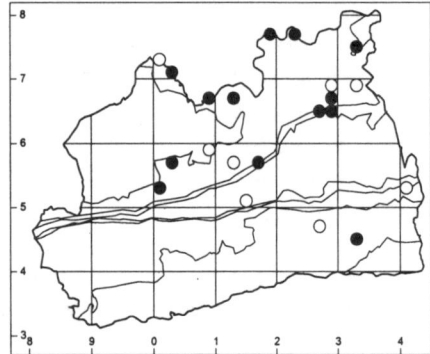

Flower visits: stinking goosefoot (Iliff, 1991), redshank, water-cress, cow parsley

Sphaerophoria scripta (Linnaeus, 1758)

Number of records: 554
Surrey Status: Ubiquitous
Flight times: April – October
Peak: August

This is a common grassland hoverfly. The lower frequency of records in south and south-west Surrey may be a feature of recorder effort, but while surveying the county I have always had the impression that this species was much less common on the wealden clays. On one occasion I observed the mating of this species on Mitcham Common (TQ2867); the male, flying over lady's bedstraw, grabbed a female in flight and the pair remained flying *in copula* for over a minute before settling together. I also possess in my collection a specimen which appears to be a gynandromorph taken on 7.7.1995 at Epsom Downs (TQ2258).

Flower visits: meadow buttercup, creeping buttercup, bulbous buttercup, lesser celandine (LP, 1950a), greater stitchwort (LP, 1950a), lesser stitchwort, sand spurrey, redshank, hawthorn (LP, 1950a), kidney vetch, black medick, wood spurge (LP, 1957), cow parsley, burnet-saxifrage, wild angelica (Uffen, 1969), wild parsnip, hogweed, upright hedge-parsley, wild carrot, common centaury, field bindweed, dodder, wild marjoram, gipsywort, butterfly-bush, heath speedwell (LP, 1950a), lady's bedstraw, heath bedstraw, creeping thistle, dandelion, cat's-ear, *Hieracium* spp., autumn hawkbit, rough hawkbit, hawkweed oxtongue, smooth hawk's-beard, common fleabane, Michaelmas-daisy, sneezewort, yarrow, scentless mayweed, common ragwort, hoary ragwort, ragwort spp.

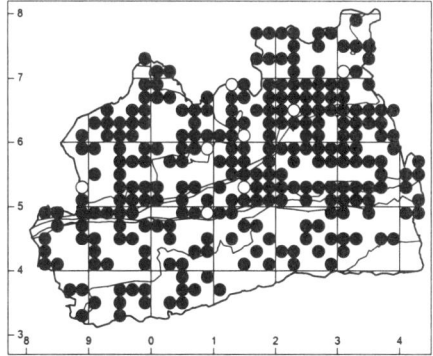

Sphaerophoria taeniata (Meigen, 1822)

Number of records: 125
Surrey Status: Common
Flight times: May – September
Peaks: May, <u>August</u>

This widespread species is fairly distinct in the field because the pale abdominal stripes are much more orange than either *S. batava* or *S. scripta*, and cover more of each tergite, thus giving the fly a brighter appearance than the aforementioned. This clue should always be followed up by identification using genital characters. *S. taeniata* is most frequently encountered in damp grasslands and is quite a strong indicator of such habitat.

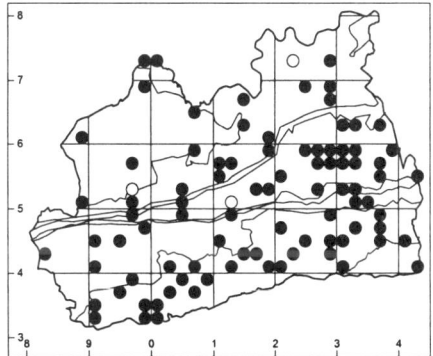

Flower visits: buttercup spp., square-stalked willowherb, burnet-saxifrage, hogweed, water mint, autumn hawkbit, common fleabane, scentless mayweed, ragwort spp., water-plantain

Sphaerophoria virgata Goeldlin de Tiefenau, 1974

Nationally Scarce
Number of records: 14
Surrey Status: Rare
Flight times: May – August
Peak: June

This is another heathland hoverfly which can be found at tormentil flowers, especially at the edges of heathland footpaths. At the moment, records of this scarce species are confined to the heaths of west Surrey.

Conservation: This species' liking of tormentil at the edge of paths and rides clearly illustrates the importance of maintaining breaks in heather cover and the presence of the heath verge habitats that accompany lightly-used tracks. Heavy use of such tracks should be avoided, as this will destroy the community of low-growing herbs and grasses.

Flower visits: heather, tormentil

RECORDS: **Hankley Common** SU8639 (1.7.1989, RKAM); **Tilford Reeds** SU8643 (13.6.1987, 28.6.1987, RKAM/GAC), (28.5.1988, RKAM); **Thursley Common** SU9041 (23.7.1967, PJC), (18.7.1990, AES), (17.7.1990, pair *in cop.*, APF); **Frith Hill** SU9058 (22.5.1988, GAC); **Wyke Common** SU9152 (14.5.1994, GAC); **Witley Common** SU9341 (16.5.1992, GAC); **Chobham Common** SU9765 (27.6.1987, RKAM); **Wisley Common** TQ0658 (29.8.1971, PJC).

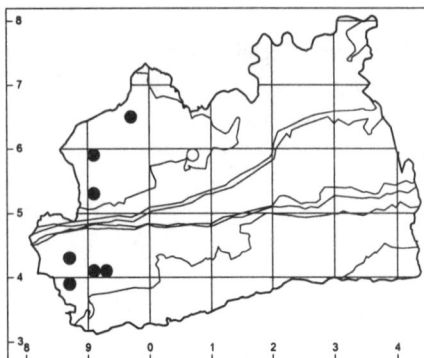

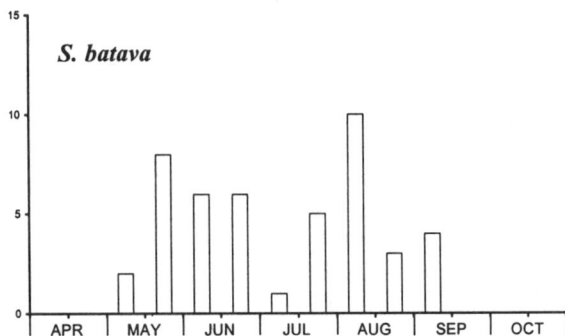

Figure 8a. **Phenology of *Sphaerophoria* species**

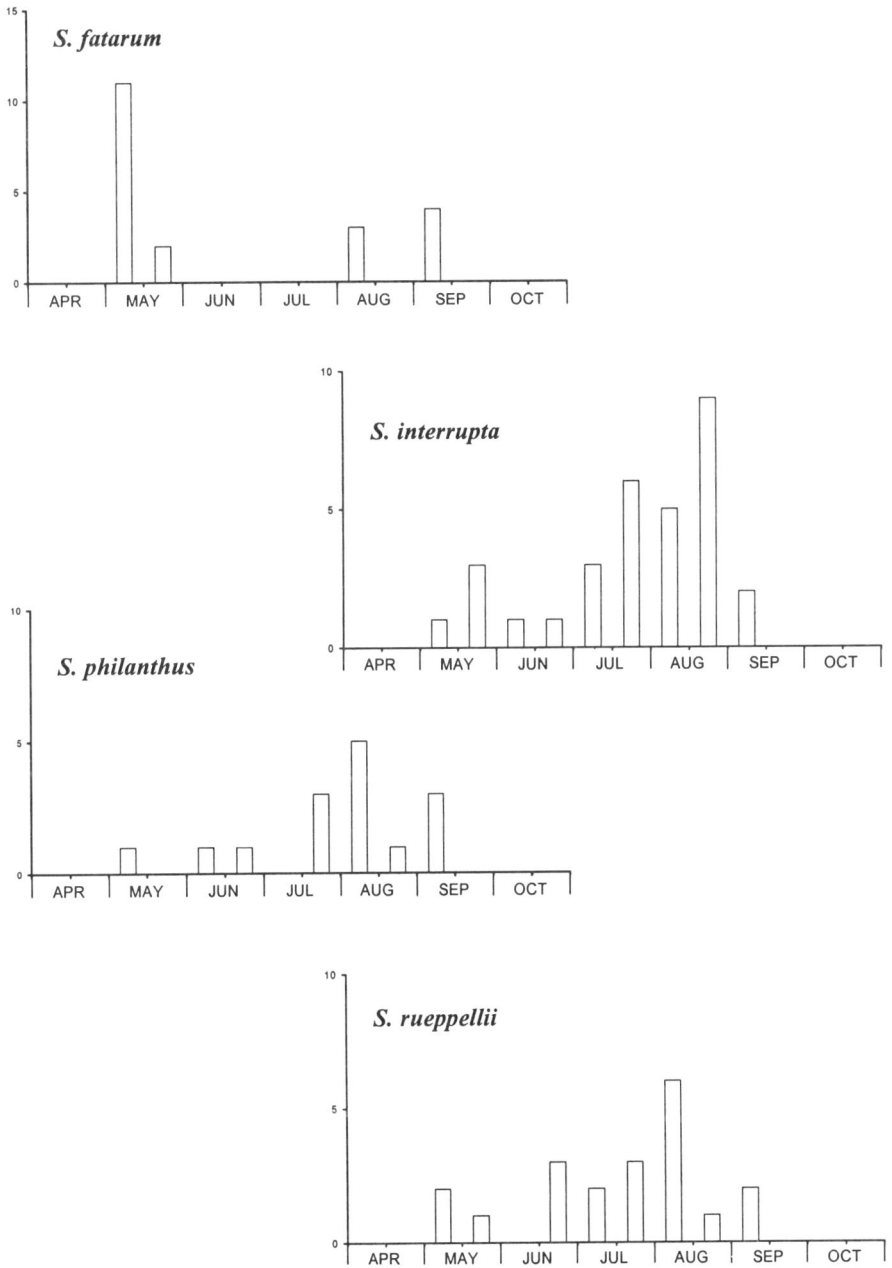

Figure 8b. Phenology of *Sphaerophoria* species

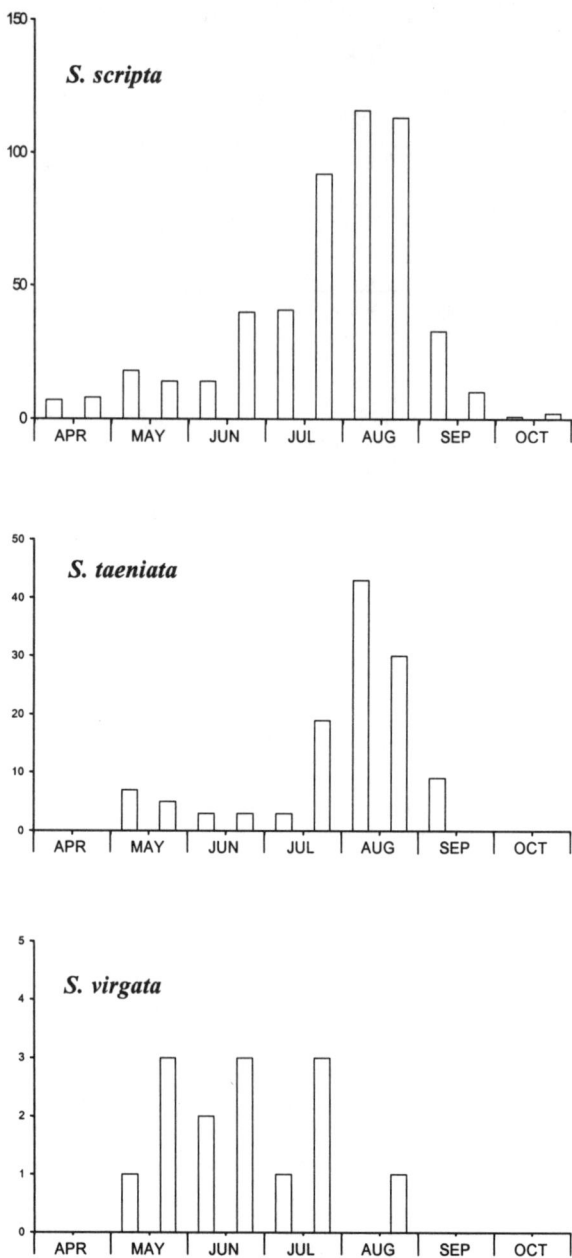

Figure 8c. Phenology of *Sphaerophoria* species

Genus *Syrphus*

These are the hoverflies most familiar to the general public, being medium-sized yellow-and-black wasp mimics; the larvae are aphid predators. All three species are very similar, so retention of a number of specimens is essential if they are to be properly recorded.

Syrphus ribesii (Linnaeus, 1758)   PLATE 2

Number of records: 901

Surrey Status: Ubiquitous

Flight times: March – October

Peaks: <u>May</u>, <u>August</u>

This is a widespread and abundant yellow-and-black hoverfly which appears to be genuinely commoner than either of the two following species. Males often frequent dappled sunlight in woodlands where they can be seen defending a shaft of sunlight; they are often responsible for the perceptible "hum" in woodland in spring and early summer. There is one record of this species suffering fungal attack (RKAM).

Flower visits: meadow buttercup, creeping buttercup (LP, 1950a), buttercup spp., greater stitchwort, sticky mouse-ear, perennial wall-rocket, heather, primrose (LP, 1950a), dog rose, wood spurge (LP, 1957), ivy, cow parsley, hemlock water-dropwort, wild angelica, wild parsnip, hogweed, upright hedge-parsley, wild carrot, field bindweed, elder, creeping thistle, common knapweed, nipplewort, cat's-ear (LP, 1956b), autumn hawkbit, perennial sow-thistle, dandelion, smooth hawk's-beard, common fleabane, oxeye daisy, common ragwort, ragwort spp.

Syrphus torvus Osten Sacken, 1875

Number of records: 180

Surrey Status: Common

Flight times: February – December

Peak: April

This is mainly a spring species and it can be very abundant in some years and almost non-existent in others. *S. torvus* is the least common of the three *Syrphus* species, but is almost certainly overlooked. There is one report of larvae taken in May 1912 from young pines on Horsell Common, and reared out to yield 30 adults between 28 and 30.5.1912 (Champion, 1912).

Flower visits: sallow (LP, 1950a), primrose (LP, 1950a), bramble, blackthorn, hawthorn (LP, 1957), Norway maple, ivy, wild angelica, hogweed, dandelion (LP, 1950a)

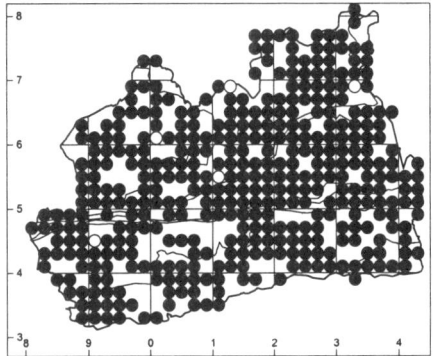

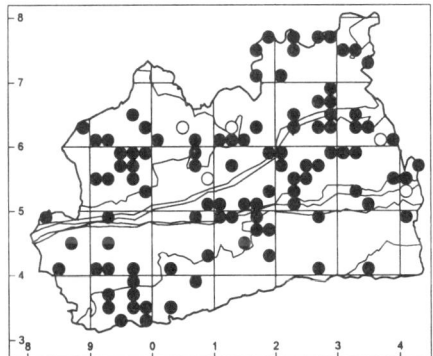

Syrphus vitripennis Meigen, 1822

Number of records: 530
Surrey Status: Ubiquitous
Flight times: April – October
Peak: August

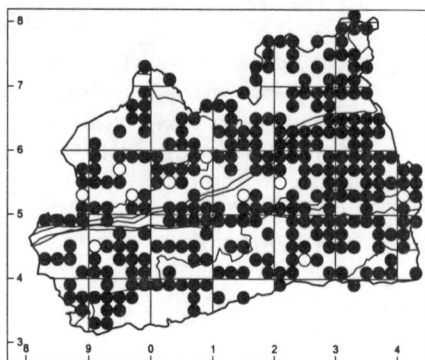

Syrphus vitripennis is certainly less common than *S. ribesii*, but is likely to have been under-recorded because members of the genus are often abundant and critical identification involves microscopic characters, therefore relying on the retention of a number of specimens, which I am often loath to do. When I made concerted attempts to record this species and *S. torvus*, the numbers proved to be considerably lower than those of *S. ribesii*, as shown in *Figure 9*. There is one record of this species suffering fungal attack (RKAM).

Flower visits: buttercup spp., perforate St John's-wort (LP, 1950a), imperforate St John's-wort, hedge mustard, meadowsweet, bramble, rowan, hawthorn, wood spurge (LP, 1957), field maple, ivy, cow parsley, burnet-saxifrage, hemlock water-dropwort, wild angelica (Uffen, 1969), wild parsnip, hogweed, upright hedge-parsley, wild carrot, field bindweed, water mint (LP, 1950a), eyebright, common marsh bedstraw (LP, 1950a), devil's-bit scabious (LP, 1941), common knapweed, cat's-ear, autumn hawkbit, bristly oxtongue, dandelion, smooth hawk's-beard, common fleabane, scentless mayweed, common ragwort, hoary ragwort, ragwort spp.

S. ribesii		*S. torvus*		*S. vitripennis*	
♂	♀	♂	♀	♂	♀
89	80	7	9	69	57

Figure 9. **Comparative numbers of *Syrphus* species recorded between 1994 and 1996**

Genus *Xanthogramma*

These are two medium to large species. Both have distinctive yellow markings and can be readily separated by the shape of the markings on the first tergite and the exact shade of the yellow colour, with *X. citrofasciatum* having a lemon-yellow and *X. pedissequum* an orange-yellow appearance. The larvae seem to be associated with ants in some way.

Xanthogramma citrofasciatum (De Geer, 1776)   PLATE 9

Number of records: 42
Surrey Status: Scarce
Flight times: April – June
Peak: May

There seems to be a strong correlation between the distribution of this striking hoverfly and the frequency of colonies of the yellow ant *Lasius flavus,* with which it is known to be associated (Speight, 1990). *X. citrofasciatum* appears to be most common on chalk downland and grass heath in south-east Surrey, but might be expected to be found elsewhere, given sufficient recording and suitable short turf with ant hills. There is one record of this species as a prey item of *Empis tessellata* (Diptera, Empididae) (LP, 1968).

Flower visits: hawthorn

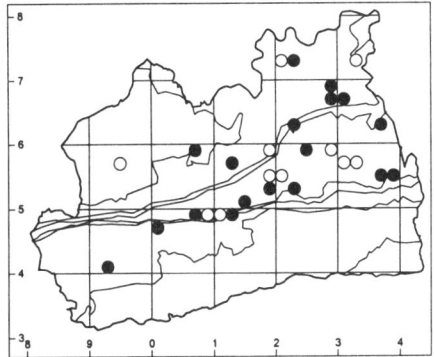

Xanthogramma pedissequum (Harris, 1776)   PLATE 9

Number of records: 221
Surrey Status: Common
Flight times: April – September
Peaks: <u>June</u>, August

This widespread species has loose affinities with grasslands and woodland rides. It is known to be associated elsewhere with the black ant *Lasius niger* (Foster, 1987), which is more tolerant of ranker vegetation than *L. flavus.* This would help to explain the more ubiquitous distribution of *X. pedissequum.*

Flower visits: creeping buttercup, wild parsnip, hogweed, field bindweed, dandelion

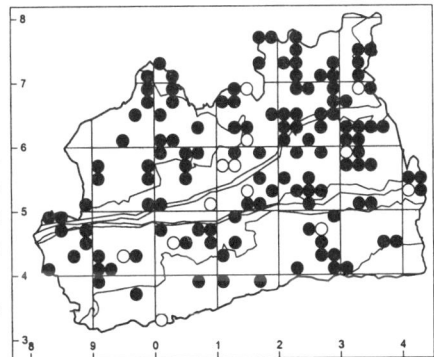

MILESIINAE

Callicerini

Genus *Callicera*

There are three species represented in Britain, only one of which occurs in Surrey; the others are confined to Scotland and East Anglia respectively. The larvae inhabit water-filled rot holes in a variety of trees. One species, *C. spinolae,* is so rare that it is only known to breed in one beech tree near Cambridge; it is on the short list of the UK Biodiversity Action Plan (DoE, 1995).

Callicera aurata (Rossi, 1790) = *Callicera aenea*: auctt., misident.   PLATE 5

Red Data Book 3
Number of records: 5
Surrey Status: Rare
Flight times: June – August
Peak: July

The larval stages of this species are known from rot holes in beech, and there is a report of larvae from a rot hole in birch in Suffolk (Perry, 1996). The latter may help to explain two records from heathland in Surrey where birch is abundant and old beech very scarce or absent. Most of the Surrey records for this species are old, and because of its narrow emergence period it is possible that it is more widespread than it appears. In flight, *C. aurata* closely resembles a bee and may therefore be overlooked; it is one of a number of species which may be more easy to find as a larva than as an adult.

Conservation: Large trees with rot holes should be retained. In some localities, such as heathland, there is the possibility that rot holes in birch are important; hence large old birches with rot holes should be identified and retained wherever possible, if considering scrub removal.

Flower visits: firethorn

RECORDS: **Windsor Great Park** SU9769 (19.6.1996, RKAM); **Wisley** (7.7.1934, GFW); **Oxshott** TQ1461 (25.7.1937, KMG); **Ashtead Common** TQ1859 (17.7.1960, SFI); **Addington** (27.8.1974, JR).

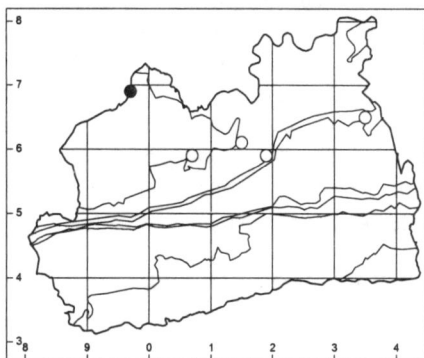

Cheilosini

The larvae of most of this tribe are plant feeders, mining roots, leaves and stems. Many are associated with thistles, but others are more specialised. Given this tribe's close associations with plants whose distribution often reflects local geology, the biogeography of this group is perhaps one of the easiest to interpret. Other species are fungal feeders, whilst a few feed on dung or sap runs. There are a number of nationally scarce species which are well represented in Surrey, and this gives considerable scope for research into the biology of many poorly-known species.

Genus *Cheilosia*

This is a large genus of black species which are often ignored by inexperienced recorders. In fact, once one becomes familiar with the features such as dust bars, facial hairs, scutellar hair colour and leg colouration, they are a nice group to study. The larvae are generally root and stem borers, and leaf miners, although there are exceptions which are associated with fungi. Many of the adults visit umbellifers (see *Figure 10*) and are best sought on plants such as hogweed, wild parsnip, wild carrot and cow parsley; some of these species are closely associated with umbellifers as larvae and perhaps others will prove to be, so records of flower visits may provide a clue to the life histories of the less well known species. Voucher specimens for the less common species are desirable and should be retained until the recorder is familiar with the genus as a whole, because some species are subtly different.

Cheilosia albipila Meigen, 1838   PLATE 15

Number of records: 38
Surrey Status: Local
Flight times: March – May
Peak: April

This species is most frequently encountered on the clays of the Low Weald and the Thames Basin, but rarely elsewhere. During this study no attempt has been made to record larvae systematically, but there are records of larvae (which are easier to find than adults) collected from marsh thistle at Great Bookham and Ashtead Commons (GAC). It is very probable that *C. albipila* will prove to be more widespread than current records of this early spring species suggest. Many records are for females ovipositing or resting on the leaf-rosettes of

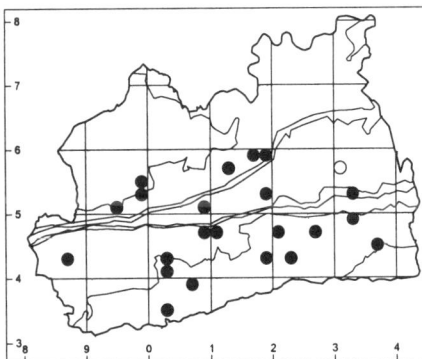

marsh thistle, but adults also have a habit of sunning themselves on dry leaves in March and may be missed unless the recorder is alert to the possibilities of insects on the ground.

Flower visits: sallow

	Ranunculaceae	Umbelliferae	Compositae	Others
C. albipila				2
C. albitarsis	27			3
C. antiqua	1			
C. barbata		5		
C. bergenstammi	2	1	8	
C. fraterna	5		2	
C. grossa			2	2
C. illustrata		37		1
C. impressa		16		1
C. lasiopa	1	1		
C. latifrons			1	
C. nigripes	1			
C. pagana	13	22	4	14
C. praecox				3
C. proxima		10	1	1
C. scutellata		17	2	
C. soror		30	1	
C. variabilis	3	6		3
C. vernalis	1	2	2	
C. vulpina		9		
Total	**54**	**156**	**23**	**30**
% records	*20.53*	*59.31*	*8.75*	*11.41*

Figure 10. Flower visit records for individual *Cheilosia* species

Cheilosia albitarsis (Meigen, 1822)

Number of records: 467
Surrey Status: Ubiquitous
Flight times: April – August
Peak: May

This common species is found at the flowers of the creeping and bulbous buttercups; I have rarely found it at meadow buttercup flowers, despite much searching, even when the other buttercup species are present and the fly is abundant. This is a species which is common for a relatively short period and is consequently under-recorded, but does appear to be absent from seemingly suitable habitat on occasions. There are two records of this species as a prey item of *Empis tessellata* (Diptera, Empididae) (LP, 1968).

Flower visits: meadow buttercup, creeping buttercup, bulbous buttercup, buttercup spp., lesser celandine, garlic mustard, silverweed (LP, 1950a), woodruff

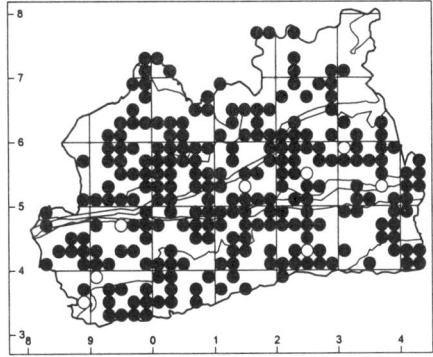

Cheilosia antiqua (Meigen, 1822)

Number of records: 6
Surrey Status: Rare
Flight times: April – May
Peak: May

The distribution of *C. antiqua* is curious as it has been found in the clay woodlands of the Weald and also on acidic gravels in London. It is associated with primrose, *Primula vulgaris*, and I can only explain the record from Putney as arising from importation with plants collected from the wild, or possibly with cultivated primulas [this species is otherwise only known in the London area from Bexley, Kent (Plant, 1986)]. It is possible that *C. antiqua* has been under-recorded because it is about the size of *C. albitarsis* and may have been overlooked when the latter was at its peak. I have rejected a small number of records of this species because males of *Melanogaster hirtella* have a small but distinct "nose" which results in specimens keying out to *C. antiqua*. All specimens of this species should therefore be retained for critical examination.

Flower visits: marsh marigold, buttercup spp.

RECORDS: **Cosford Mill** SU9139 (11.5.1969, PJC); near **Chiddingfold** SU9634 (30.4.1994, RKAM); **Tuesley** SU9642 (25.5.1987, GAC); **Compton Common** SU9647 (16.5.1992, RKAM); **Vann Copse** SU9837 (25.5.1986, GAC); **Putney Heath** TQ2374 (30.5.1992, RKAM).

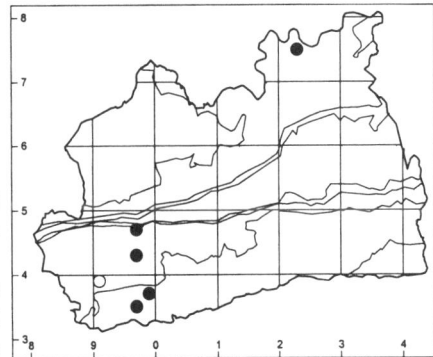

Cheilosia barbata Loew, 1857

Nationally Scarce
Number of records: 58
Surrey Status: Local
Flight times: June – August
Peak: July/August

The first published British record of this hoverfly was from Leith Hill in June 1868 (Verrall, 1901), and until the 20th century this was one of just two British records. Despite its apparent national scarcity, *C. barbata* is widely distributed in Surrey. The majority of records are from the Chalk, but a substantial number come from the adjacent clays of the Low Weald and the Thames Basin. Further records come from the Greensand in the vicinity of Friday Street. Such a wide range of geology makes interpretation of the species' ecology very difficult, but it does appear to be a woodland hoverfly which is commonly found at hogweed and would seem to favour basic soils. Its larval food-plant is unknown.

Flower visits: wild parsnip, hogweed

Figure 11. **Phenology of *Cheilosia barbata***

RECORDS: **Park Copse** SU9137 (7.8.1987, RKAM); **Hog's Back** SU9248 (15.8.1987, RKAM); **Frillinghurst Woods** SU9334 (7.8.1987, RKAM); near **Willey Green** SU9450 (11.7.1987, RKAM); **Littlefield Common** SU9552 (29.7.1989, RKAM); **Shalford** SU9947 (15.7.1989, GAC); **Shalford Common** TQ0046 (3.8.1995, RKAM); **Merrow Common** TQ0251 (5.7.1987, RKAM); **Run Common** TQ0341 (16.8.1987, RKAM); **Merrow Downs** TQ0349 (29.8.1988, GAC); **Newlands Corner** TQ0450 (7.7.1993, JRD), TQ0449 (29.8.1988, GAC); **Hurtwood** TQ0644 (22.7.1995, RKAM); **Albury Heath** TQ0646 (16.8.1987, RKAM); **Gomshall** TQ0847 (5.7.1988, GAC); **Hackhurst Downs** TQ0948 (27.8.1966, PJC), (2.7.1966, SBRC); **Oldlands Wood** TQ1051 (7.7.1988, GAC); **Friday Street** TQ1245 (17.7.1985, RKAM); **Wotton** TQ1245 (19.7.1964, AWJ); **Great Bookham Common** (no data, LP), TQ1256 (2.7.1986, 22.6.1988, 20.7.1988, GAC), (17.6.1987, RKAM/GAC); **Leith Hill** (25.6.1868, GHV); **Westcott Downs** TQ1349 (16.7.1985, RKAM), (22.7.1989, GAC); **Great Oaks** TQ1562 (9.8.1987, RKAM); **Box Hill** (1867, GHV), TQ1852 (6.8.1989, RKAM), (25.6.1989, AJH), TQ1851 (21.8.1977, AES), (14.8.1987, RKAM), TQ1751 (26.7.1953, GCDG); **Ashtead Common** TQ1859 (22.6.1986, GAC), (9.8.1986, RKAM/GAC), (26.7.1987, RKAM); **Headley Heath** TQ2053 (20.7.1989, RKAM); **Walton on the Hill** TQ2155 (6.8.1989, RKAM); **Colley Hill** TQ2452 (9.8.1989, RKAM); **Burgh Heath** TQ2457

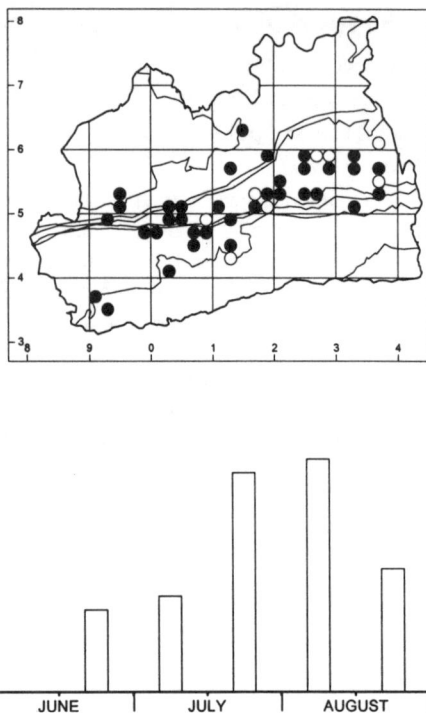

(17.7.1995, RKAM); **Ruffett Wood** TQ2558 (8.8.1989, RKAM); **Chipstead** (8.8.1953, LP); **Upper Gatton** TQ2753 (2.8.1987, RKAM/GAC); **Hooley Downs** TQ2956 (23.7.1995, RKAM); **Coulsdon Common** TQ3257 (28.7.1995, RDH); **Coulsdon** (17.7.1949, LP); **Woldingham** (21.7.1950, SW); **Bletchingley Sandpit** TQ3350 (6.8.1995, RKAM); **Riddlesdown Quarry** TQ3359 (19.8.1987, GAC); **Selsdon** (3.8.1929 - 11.8.1929, RLC); **Tandridge Hill** TQ3653 (17.7.1994, RKAM); **Worms Heath** TQ3757 (22.8.1987, RKAM/GAC).

Cheilosia bergenstammi Becker, 1894

Number of records: 167
Surrey Status: Common
Flight times: April – September
Peaks: <u>May</u>, August

This is a common species which peaks in spring and again in midsummer. The spring generation is most frequently found at dandelion whilst the summer generation is mostly found at ragwort flowers. The larvae are associated with ragwort, a plant which is widespread in Surrey, yet the distribution of the adult is distinctly clumped around relatively dry habitats such as heathland and downland. The absence of this species from the London area is curious, as ragwort can be common on sites such as Mitcham Common where *C. bergenstammi* has never been recorded, despite intensive survey.

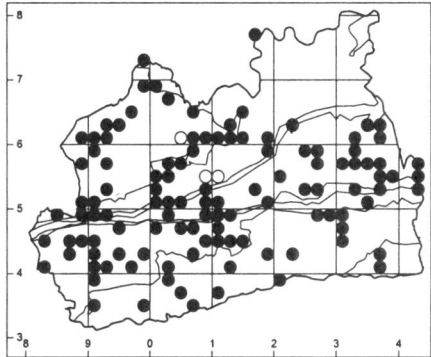

Flower visits: buttercup spp., wood spurge (LP, 1957), upright hedge-parsley, dandelion, common fleabane, common ragwort, ragwort spp.

Cheilosia carbonaria Egger, 1860

Nationally Scarce
Number of records: 15
Surrey Status: Rare
Flight times: May – September
Peaks: <u>May/June</u>, August/September

Many records of this hoverfly are from old woodland such as Hammonds Copse and Tugley Wood, and most are from clays, especially those north of the North Downs. Geology is not the only key to this species' ecology, however, and there are a number of unusual records such as those from Windlesham and Mitcham Common. It is possible that this species is overlooked amongst the plethora of *C. albitarsis* in the spring.

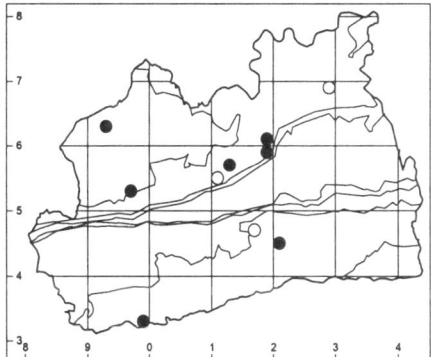

Flower visits: hawthorn (LP, 1950a)
RECORDS: **Windlesham** SU9263 (29.7.1989, RKAM); **Whitmoor Common** SU9753

(29.5.1994, RKAM); **Tugley Wood** SU9833 (10.6.1989, RKAM); **Effingham** (11.6.1941, NHML); **Great Bookham Common** (10.8.1947, NHML), (12.9.1954, LP), TQ1256 (14.5.1995, GAC); **Holmwood** (7.6.1964, AWJ); **Ashtead Common** TQ1859 (26.5.1973, PJC), (11.6.1986, GAC); **Epsom Common** TQ1860 (7.9.1985 RKM/GAC), (11.6.1987, RKAM); **Mitcham Common** (16.8.1959, LP); **Hammonds Copse** TQ2144 (21.8.1993, RKAM).

[*Cheilosia cynocephala* Loew, 1840

Nationally Scarce

Number of records: 2

Surrey Status: Unconfirmed

A recent record from Mitcham Common involves a site where musk thistle, *Carduus nutans*, the presumed larval foodplant, does not occur (Lousley, 1976). There is also a record from Dulwich on the Hoverfly Recording Scheme database; this too is an unlikely site. I have not been able to examine these specimens and have therefore listed this species as unconfirmed from Surrey although further survey for this species, visiting sites known to support musk thistle, may be worthwhile.]

Cheilosia fraterna (Meigen, 1830)

Number of records: 42

Surrey Status: Local

Flight times: April – August

Peaks: May, August

The distribution of *C. fraterna* suggests that it is a scarce species in Surrey, which is possibly true. It may, however, have been overlooked because it often occurs at the same time and in the same places as *C. bergenstammi*, and may be missed when the latter is abundant. The larvae tunnel the stem and rosette of marsh thistle (Rotheray, 1993), a widely distributed plant which is commonest on clay. This may help to explain the distribution of this hoverfly which is apparently absent from the Chalk and concentrated on the wet heaths of west Surrey and on the Low Weald.

Flower visits: meadow buttercup, creeping buttercup (LP, 1950a), buttercup spp., silverweed (LP, 1950a), common fleabane, common ragwort

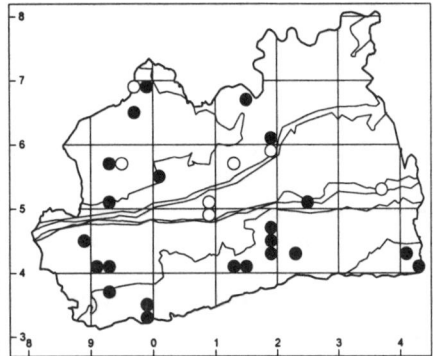

Cheilosia griseiventris Loew, 1857

Number of records: 1

Surrey Status: Rare

Flight times: August

It is likely that *C. griseiventris* will prove to be slightly more common when recorders are confident about assigning specimens to this species. However, neither this nor the very similar *Cheilosia latifrons* (= *intonsa*) is common in Surrey. The only record from Surrey is that of a male from a woodland edge on clay adjacent to a bean field.

RECORDS: near **Hungry Hill** TQ0555 (23.8.1992, RKAM).

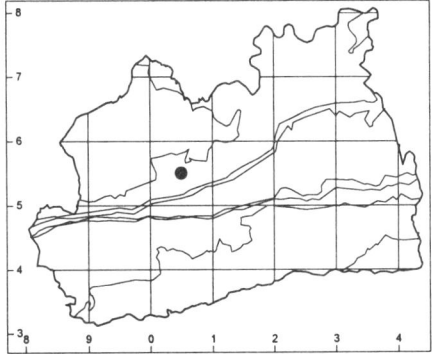

Cheilosia grossa (Fallén, 1817)   PLATE 15

Number of records: 19

Surrey Status: Rare

Flight times: March – April

Peak: April

This appears to be a very local species which is possibly more frequent towards the London suburbs, in contrast to *C. albipila* which is mainly found west and south of the Chalk. Rotheray (1993) quotes spear thistle, *Cirsium vulgare*, as a larval foodplant and it has been found in other thistles, especially marsh thistle (Rotheray, *pers. comm.*); this helps to explain why it is not found in the same places as *C. albipila*. *C. grossa* is probably significantly under-recorded as it flies in the early spring when conditions are often unsuitable for recording; the larvae are easier to find, but have not been recorded systematically. Males are more commonly encountered than females and may be observed defending a territory in open spaces such as over footpaths. They frequently fly high up and may also be missed for this reason. Adults are often recorded from sallow catkins.

Flower visits: sallow, colt's-foot

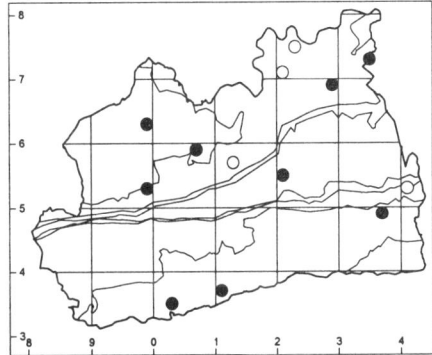

Cheilosia illustrata (Harris, 1780)   PLATE 1

Number of records: 324
Surrey Status: Common
Flight times: May – September
Peak: July

This is a widespread and common species whose larvae are associated with hogweed (G. Rotheray, *pers. comm.*). Adults are most frequently found at hogweed flowers (90% of records) in woodland rides.

Flower visits: buttercup spp., ground-elder, wild parsnip, hogweed

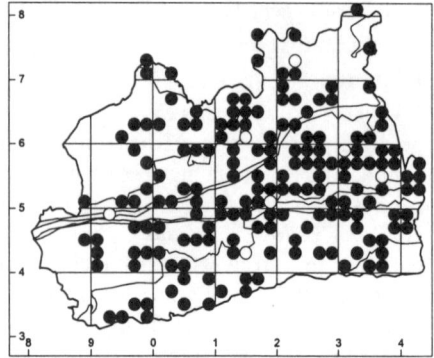

Cheilosia impressa Loew, 1840

Number of records: 123
Surrey Status: Common
Flight times: April – September
Peak: August

This locally common hoverfly can be found at a wide variety of umbels in and around woods. It would appear to be most common on the Chalk and on the Greensand and Wealden clays of west Surrey. This is a curious distribution which cannot be explained at this stage. The distinctive yellow wing-bases of the female are characteristic, even in the field, but small *C. albitarsis* could be mistaken for this species, so it is generally wise to retain specimens for confirmation.

Flower visits: water-cress, cow parsley, burnet-saxifrage, fool's water-cress, wild parsnip, hogweed, wild carrot

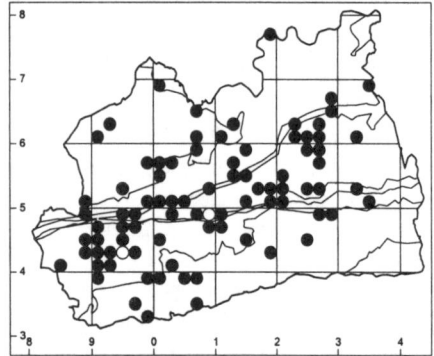

Cheilosia lasiopa Kowarz, 1885 = *Cheilosia honesta*: Verrall, 1901, misident.

Number of records: 55
Surrey Status: Local
Flight times: April – June
Peak: May

Cheilosia lasiopa seems to occur most frequently in woodland and at woodland edge. The maps suggest that this species is associated with drier habitats on chalk and sand, and that it is absent from the clay of the Low Weald. It has a relatively short emergence period and may consequently be under-recorded.

Flower visits: buttercup spp., cow parsley

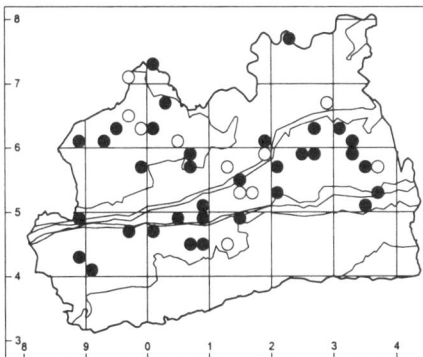

Cheilosia latifrons (Zetterstedt, 1843) = *Cheilosia intonsa* Loew, 1857

Number of records: 8
Surrey Status: Rare
Flight times: May – August
Peak: August

There are very few records of this hoverfly, usually represented by single specimens from dry sites on sand; most were at yellow composites, but one was at yarrow on Wimbledon Common. It is a smaller, dumpier species than *C. griseiventris* and, once known, males are easily recognised; females remain more problematic (S. Falk, *pers. comm.*).

Although the sample is very small, there would seem to be two broods; there is one record from May, none in June and the majority in July/August.

Flower visits: yarrow

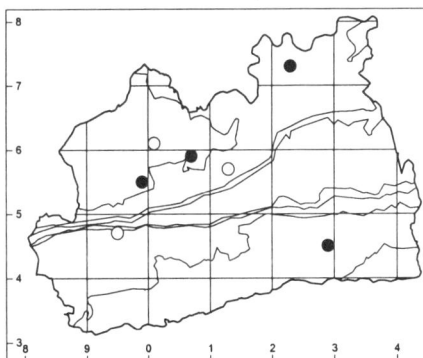

RECORDS: **Puttenham Golf Course** (17.5.1929, CD); **Smarts Heath** SU9855 (11.7.1987, GAC); **Horsell Common** TQ0160 (1980 - 1981, AJH); **Wisley RHS Gardens** TQ0658 (1982, AJH); **Great Bookham Common** (no data, LP, 1960); **Horsley** (1957, FB); **Wimbledon Common** TQ2272 (28.8.1985, RKAM); **Horley** TQ2844 (28.8.1993, RKAM).

Cheilosia longula (Zetterstedt, 1838)

Number of records: 19
Surrey Status: Rare
Flight times: June – October
Peak: September

The larvae of this species are associated with *Boletus* fungi, which are widespread on heathland, and this seems to make *C. longula* a good heathland indicator. There is a rearing record from Thursley Common (SU9041) of larvae in *Boletus luteus* (PJC). Most records are from late in the recording season and it is probable that *C. longula* will prove to be more common than current records suggest.

RECORDS: **Hindhead Common** SU8936 (5.8.1991, SJG); **Thursley Common** SU9041 (28.8.1967, 8.9.1970, PJC); **Puttenham Common** SU9045 (6.8.1988, GAC); **Windlesham** SU9263 (29.7.1989, RKAM); **Witley Common** SU9341 (21.8.1992, PJH); **Whitmoor Common** SU9853 (25.6.1988, GAC), (11.9.1988, RKAM); **Hoe Stream, Milford** SU9956 (1984 - 1989, AJH); **Horsell Common** TQ0160 (11.9.1988, RKAM); **Wisley Common** TQ0658 (3.9.1978, AES), (14.8.1987, RKAM); **Wisley RHS Gardens** TQ0658 (1982, AJH); **Hurtwood** TQ0843 (4.9.1988, RKAM/GAC); **White Downs** TQ1148 (1982, AJH); **Oxshott Heath** (2.9.1951, 19.10.1964, LP); **Limpsfield Common** (pre-1941, LP).

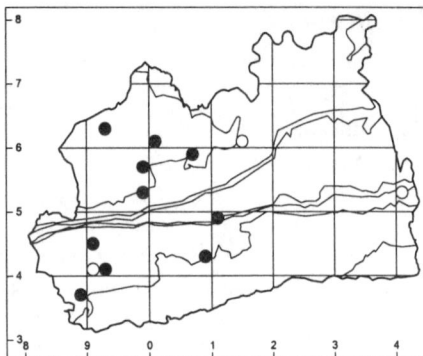

Cheilosia mutabilis (Fallén, 1817)

Nationally Scarce
Number of records: 3
Surrey Status: Rare
Flight times: May – October

This is a very poorly known hoverfly with just a single modern record. Critical examination of *C. albitarsis* often yields specimens with dark halteres, but the shape and size are wrong for *C. mutabilis* which is a small narrow-bodied hoverfly. I have examined the specimen from Ashtead Common and, although it seems to fit this species, it does have yellow on the middle tibia and there is little evidence of black on the halteres which makes it a little suspect.

RECORDS: **Oxshott Heath** (19.10.1964, LP); **Ashtead Common** TQ1859 (2.7.1986, GAC); **Headley Heath** (3.5.1925, OWR).

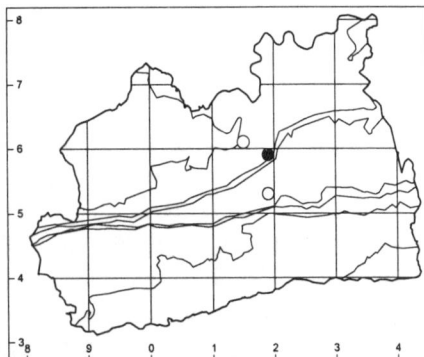

Cannon Hill Common

Epistrophe grossulariae

Leucozona glaucia

Cheilosia illustrata

Eristalis pertinax

Woodland

PLATE 1

Melangyna compositarum/labiatarum

Leucozona lucorum

Epistrophe eligans

Syrphus ribesii

Ramsons at Newdigate

Portevinia maculata

Woodland

PLATE 2

Pirbright Ranges

Didea fasciata

Dasysyrphus venustus

Dasysyrphus tricinctus

Coniferous woodland

PLATE 3

Stump in Margery Wood

Xylota sylvarum

Criorhina asilica (in cop.)

Criorhina floccosa

Caliprobola speciosa

Dead wood

PLATE 4

Beech tree at Box Hill

Pocota personata

Callicera aurata

Dead wood

PLATE 5

Rot hole in oak, Richmond Park

Brachypalpus laphriformis

Water cavity in root plate

Myathropa florea

Dead wood

PLATE 6

Fallen beech at Margery Wood

Sap run on Bookham Common

Brachyopa insensilis larva

Brachyopa insensilis

Dead wood – Brachyopa habitat

PLATE 7

Box Hill

Doros profuges

Doros profuges

Microdon devius

Platycheirus manicatus

Chalk grassland

PLATE 8

Xanthogramma citrofasciatum

Xanthogramma pedissequum

Chrysotoxum festivum

Chrysotoxum cautum

Grassland species

PLATE 9

MOD Ranges

Sericomyia lappona

Sericomyia silentis

Heathland

PLATE 10

Pine stump at Pirbright Ranges

Microdon analis puparium and *Lasius niger*

Microdon analis puparium

Microdon analis

Microdon analis – life history

PLATE 11

Alder carr at Gomshall

Tuesley

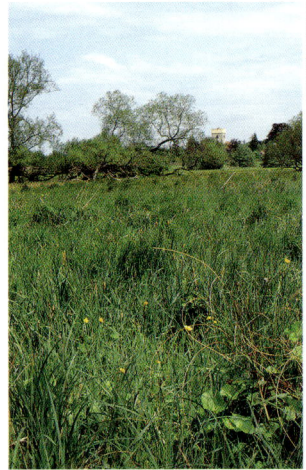

Water meadows at Send Grove

Neoascia spp.

Wetlands

PLATE 12

Helophilus hybridus

Anasimyia contracta ovipositing on *Typha* leaf

Parhelophilus spp.

Tropidia scita

Wetlands

PLATE 13

Merodon equestris – larva in bulb

Merodon equestris form equestris

Merodon equestris form narcissi

Merodon equestris form equestris

Merodon equestris form narcissi

Merodon equestris – colour forms

PLATE 14

Cheilosia albipila larva

Thistle affected by *Cheilosia* larvae

Cheilosia grossa puparium

Eristalis interruptus courtship

Meligramma euchromum puparium

Life history

PLATE 15

Volucella bombylans

Volucella bombylans

Volucella pellucens

Volucella inanis

Volucella zonaria

Spectacular species

PLATE 16

Cheilosia nebulosa Verrall, 1871

Red Data Book 3
Number of records: 7
Surrey Status: Rare
Flight times: April – May
Peak: April

Records of this species are widespread and mainly of single individuals. It appears to be a woodland species which occurs most frequently in the Chiddingfold area, but other habitat preferences are unclear. The Botany Bay records are of several males hovering at head height, something which I have only witnessed for *C. praecox* (q.v.), and a female crawling over low vegetation (G. A. Collins, *pers. comm.*).

RECORDS: **Botany Bay** SU9734 (8.4.1989, GAC), SU9834 (30.4.1986, GAC); **Great Bookham Common** (14.4.1946, CNC), TQ1256 (10.4.1988, RBH); **Headley Heath** TQ1953 (1.4.1995, GAC); **Ruffett Wood** TQ2558 (3.5.1993, GAC).

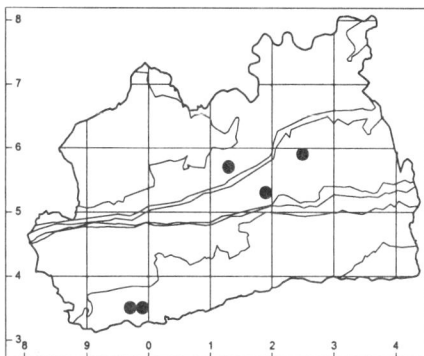

Cheilosia nigripes (Meigen, 1822)

Red Data Book 3
Number of records: 24
Surrey Status: Scarce
Flight times: May – June
Peak: June

This species flies in late spring and often visits yellow flowers such as buttercups. It occurs in herb-rich rides and glades in woodland on the Chalk, and although usually found in low numbers it can occasionally be abundant. It is likely that this species will prove to be more widespread than current records suggest, but it is clearly confined to the Chalk. Adults, especially males, are frequently found sunning on leaves, behaviour which I have also observed in *C. vicina* in Scotland. This is one of the many spring *Cheilosia* which are very similar, and consequently it may be overlooked, especially where *C. albitarsis* is common.

Conservation: Maintenance of open, sunny rides and glades is an important part of woodland management from which this species will benefit.

Flower visits: buttercup spp.

RECORDS: **Gomshall** TQ0847 (1.6.1992, JRD); **Hackhurst Downs** TQ0948 (21.6.1986, RKAM); **White Downs** TQ1148 (15.6.1985, PJH), **White Hill** TQ1252 (11.6.1994, RKAM); **Ranmore** TQ1250 (2.6.1992, JSD), TQ1451 (16.6.1988, RKAM); **Box Hill** TQ1751 (11.6.1978, AES), (27.5.1980, SJF), TQ1752 (12/13.6.1979, AES), (26.5.1987, RKAM), TQ1851 (12.6.1979, AES), (1991 - 1992, JSD) ; **Headley Heath**

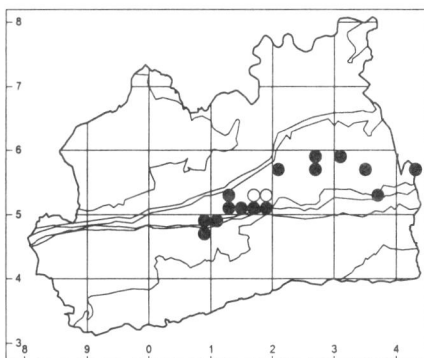

(11.6.1965, LP); near **Fourfield Close** TQ2056 (5.5.1995, RKAM); **Chipstead Bottom** TQ2657 (3.6.1987, 7.6.1988, GAC); **Banstead Park Downs** TQ2658 (10.6.1994, RDH); **Dollypers Hill** TQ3158 (20.5.1989, PLTB); **Caterham** TQ3457 (19.5.1990, RKAM); near **Godstone** TQ3652 (1993, CWP); **Tatsfield** TQ4256 (21.5.1987, 12.6.1987, GAC).

Cheilosia pagana (Meigen, 1822)

Number of records: 588
Surrey Status: Ubiquitous
Flight times: March – October
Peaks: May, <u>August</u>

This common hoverfly is likely to be found in almost any locality in Surrey. It is common at the flowers of lesser celandine in the spring and can be found at a variety of umbellifers and yellow composites later in the year. The larvae have been found to be associated with the decaying roots of cow parsley (Stubbs & Falk, 1983). *C. pagana* is a very variable species which exhibits strong seasonal dimorphism.

Flower visits: marsh marigold, creeping buttercup, buttercup spp., lesser spearwort, lesser celandine, greater stitchwort, lesser stitchwort, water chickweed, Japanese knotweed, garlic mustard, perennial wall-rocket, creeping cinquefoil, bird cherry, rowan, wood spurge (LP, 1957), cow parsley, burnet-saxifrage, ground-elder, wild angelica, wild parsnip, hogweed, upright hedge-parsley, wild carrot, forget-me-not, germander speedwell, perennial sow-thistle, dandelion, ragwort spp., colt's-foot, water-plantain

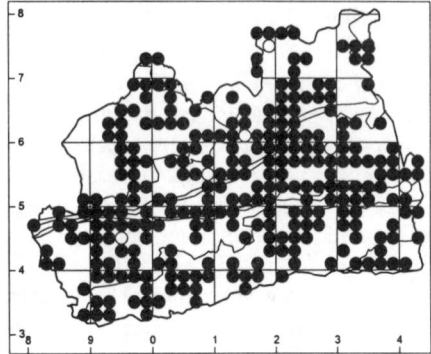

Cheilosia praecox (Zetterstedt, 1843)

Number of records: 38
Surrey Status: Local
Flight times: April – June
Peak: May

This is a small narrow species which may easily be confused with *Platycheirus ambiguus* in the field. It is mainly associated with dry well-drained soils, principally in west Surrey where I have found females flying low over short turf on downland and free-draining sandy soil. Females are reported to oviposit on mouse-ear hawkweed (Claussen, 1980), and it is possible that they also develop in the rosettes of other yellow composites such as cat's-ear and hawkbit which are common in such situations. Males have been observed flying in small swarms in sunlit situations (Morris, 1991) and at scrub edge. Although most specimens have orange antennae, I have seen some with dull brown antennae.

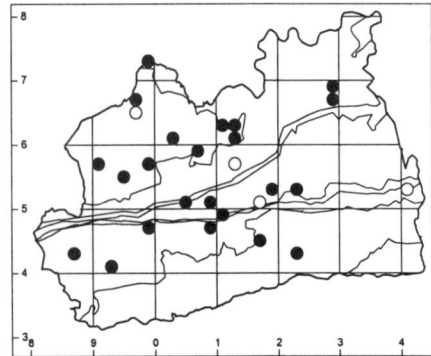

Flower visits: creeping willow, wild cherry

Cheilosia proxima (Zetterstedt, 1843) = Species D & E of Stubbs & Falk, 1983

Number of records: 112
Surrey Status: Common
Flight times: April – September
Peaks: May, <u>August</u>

This hoverfly, whose larvae are known to be associated with marsh thistle (G. Rotheray, *pers. comm.*), is widely distributed across a range of soil types, suggesting fairly catholic preferences and perhaps an association with a wider range of thistles than is currently known. It would appear to be scarce or absent from much of inner London, and parts of outer London that are relatively well recorded, which is odd because many of the inner London sites support a range of thistles and might be expected to yield *C. proxima*. The second brood is often common at umbels and is much commoner than the first brood, which visits a range of woodland-edge umbellifers. This remains a difficult hoverfly to identify with certainty and it is important to retain all specimens of this species until the recorder is also familiar with *C. velutina*.

Flower visits: cow parsley, burnet-saxifrage, hemlock water-dropwort, wild angelica, wild parsnip, hogweed, upright hedge-parsley, common fleabane, water-plantain

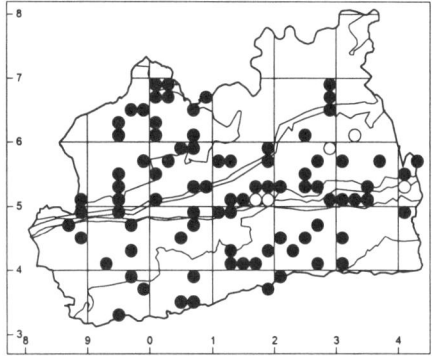

Cheilosia scutellata (Fallén, 1817)

Number of records: 128
Surrey Status: Local
Flight times: June – September
Peak: August

This is a widely distributed woodland hoverfly whose larvae develop in *Boletus* fungi; there are two rearing records, from *Boletus luteus* from Thursley Common, SU9041 (PJC), and from *B. scaber* from Chobham Common, SU9863 (PJC). Adults are most often found at umbellifer flowers, most frequently in woodlands on the Chalk and on the heathlands of north-west Surrey. There would appear to be a strong correlation with better-drained soils as this species is curiously absent from much of the Low Weald. Records from the inner London area are confined to Wimbledon Common and Kew Gardens, both of which have some woodland and are on well-drained soils.

Flower visits: ivy, wild angelica (Uffen, 1969), wild parsnip, hogweed, upright hedge-parsley, wild carrot, Canadian goldenrod, common ragwort

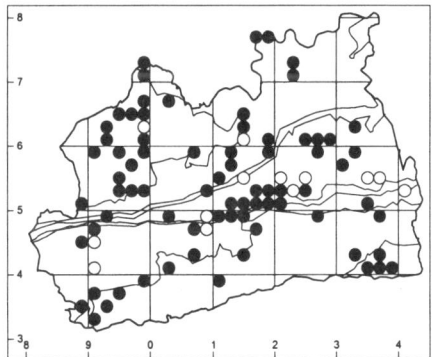

Cheilosia semifasciata Becker, 1894

Red Data Book 3
Number of records: 1
Surrey Status: Rare
Flight times: April

Most records of this hoverfly are from Wales, with a small colony in Pamber Forest (Hants) being the nearest known to Surrey. The larvae mine the leaves of orpine, *Sedum telephium*, and navelwort, *Umbilicus rupestris*. As this species is not known to be a migrant, it would appear that *C. semifasciata* is resident in Surrey on account of a male taken at Hurtwood. This is a very significant extension of the known range and indicates that searching for the larvae of this rare species may be worthwhile. The adults, which fly early in the year, are not dissimilar to *C. vernalis* and may therefore be overlooked; the male I possess has slight dusting on the tergites and the face is much more produced than *C. vernalis*, making it distinctively different when examined critically. Larval mines can be extensive and examples of the host plant riddled with leaf mines should be subject to careful scrutiny.

RECORDS: **Hurtwood** TQ0644 (14.4.1995, RKAM).

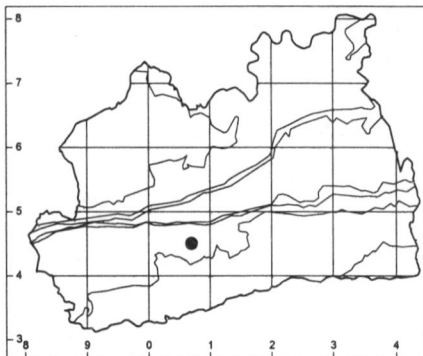

Cheilosia soror (Zetterstedt, 1843)

Nationally Scarce
Number of records: 96
Surrey Status: Local
Flight times: June – September
Peak: August

The distribution of *C. soror* mainly follows the Chalk, but there are also scattered records from the clays that overlie the Chalk to the north and south of the North Downs, including one surprising record from the Chiddingfold woods. There are also two particularly unusual records from the environs of London, one from Nunhead Cemetery and another from Richmond. These may arise from the influences of cement or lime which may make the site sufficiently alkaline to support the fungi (possibly truffles – Stubbs & Falk, 1983) on which the larvae of *C. soror* are believed to live. *C. soror* seems to confine its flower visits almost entirely to umbellifers, favouring hogweed, wild parsnip and wild carrot flowers, but it can also be found at other umbellifers. Where found, it is quite likely to be the commonest species. On one occasion I took this species at an unidentified yellow composite (? *Leontodon* spp.). The females are quite distinctive with orange antennae, bare eyes, longish body and slightly steely appearance. One useful diagnostic character of female

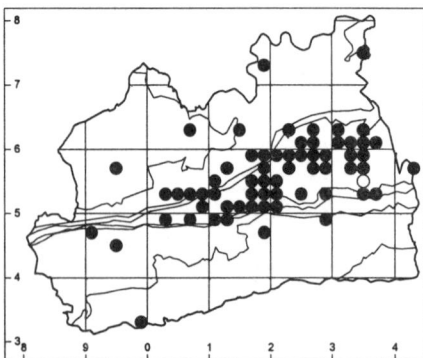

C. soror is the presence of distinctive pits on the antennae which are not present on related species with orange antennae and may be indicative of the suggested association with subterranean fungi, which would rely on refined chemosensory organs. This species is often much larger than others of the genus, so large specimens of *Cheilosia* from calcareous locations with orange antennae should always

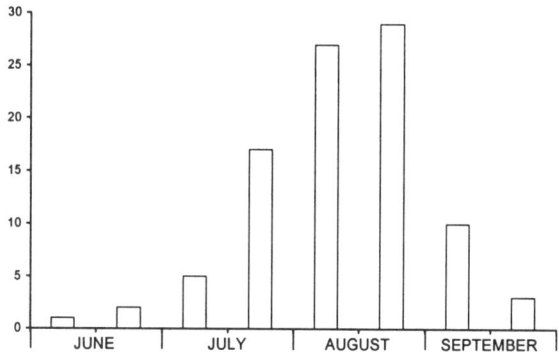

Figure 12. Phenology of *Cheilosia soror*

be retained for critical examination. Males are less distinctive and are more regularly recognised after capture. Voucher specimens should always be retained.

Flower visits: rough chervil, burnet-saxifrage, ground-elder, wild angelica, wild parsnip, hogweed, upright hedge-parsley, wild carrot

RECORDS: **Totford Lane Alder Wood** SU9147 (15.8.1992, RKAM); **Eashing** SU9444 (30.7.1994, RKAM); **Basingstoke Canal** SU9457 (16.8.1996, RKAM); **Tugley Wood** SU9833 (18.7.1993, RKAM); **Burpham Woods** TQ0252 (29.7.1994, RKAM); **West Clandon** TQ0553 (29.7.1994, RKAM); **Weybridge** TQ0663 (2.8.1997, RDH); **West Horsley** TQ0753 (12.8.1995, RKAM); **Merrow Downs** TQ0349 29.8.1988, RKAM); **Netley Downs** TQ0748 (26.7.1991, 14.8.1991, JRD); **The Sheepleas** TQ0851 (26.9.1987), TQ0852 (11.8.1991, AJH), (29.7.1994, RKAM); **East Horsley** TQ1052 (29.7.1994, RKAM); **Effingham Junction** TQ1055 (29.7.1994, RKAM/GAC); **White Downs** TQ1148 (29.8.1988, RKAM/GAC), TQ1249 (1992, JSD); **Ranmore** TQ1250 (27.8.1994, RKAM), TQ1551 (3.8.1995, RKAM); **Great Bookham Common** TQ1256 (5.8.1987, 22.6.1988, GAC), TQ1356 (7.9.1985, RKAM/GAC); **Great Oaks** TQ1562 (18.9.1988, RKAM); **Fetcham** TQ1655 (27.8.1994, RKAM); **Ashtead Common** (3.8.1947, CNC), TQ1659 (3.8.1994, PJH), TQ1859 (29.8.1985, RKAM); **Juniper Hall** TQ1751 (17.8.1979, SJF); **Box Hill** TQ1751 (14.8.1985, RKAM), TQ1752 (1.9.1991, RKAM), TQ1851 (21.8.1977, AES), TQ1852 (6.8.1989, 4.9.1994, RKAM); **Petersham Common** TQ1873 (15.7.1995, RKAM); **Roothill** TQ1947 (17.8.1985, RKAM); **Nower Wood** TQ1954 (27.8.1994, RKAM); **Stane Street** TQ1956 (10.7.1989, RKAM); **Epsom Downs** TQ2058 (19.8.1995, RKAM), TQ2258 (17.7.1995, RKAM); **Betchworth** TQ2151 (14.8.1994, RKAM); **Pebblecombe** TQ2152 (19.8.1995, RKAM); **Walton on the Hill** TQ2155 (6.8.1989, 27.8.1994, RKAM); **Walton Downs** TQ2257 (12.9.1985, GAC); **Nonsuch Park** TQ2362 (29.8.1993, PRH); **Colley Hill** TQ2452 (26.8.1994, RKAM); **Ruffett Wood** TQ2558 (8.8.1989, 26.8.1994, RKAM); **Banstead Downs** TQ2560 (12.9.1985, GAC), TQ2561 (10.8.1986, RKAM); **Chipstead Valley** TQ2657 (26.8.1994, RKAM); **Banstead Park Downs** TQ2658 (16.6.1994, 5.7.1994, 2.8.1994, 7.9.1994, 22.7.1995, RDH); **Banstead Wood** TQ2758 (28.8.1994, RKAM); **Oaks Park** TQ2761 (17.7.1995, RKAM), TQ2762 (26.8.1994, RKAM); **Redhill** TQ2848 (6.8.1994,

RKAM); **Boars Green** TQ2853, (6.8.1994, RKAM); **Hooley Downs** TQ2956
(23.7.1995, RKAM); **Farthing Downs** TQ2958 (8.8.1995, RDH); **Roundshaw** TQ3062
(23.7.1995, RKAM); **Higher Drive Recreation Ground** TQ3160 (28.8.1994, RKAM);
Coulsdon Common TQ3257 (28.7.1995, RDH); **Riddlesdown** TQ3260 (13.6.1988,
GAC), (25.7.1996, RDH), TQ3359 (9.8.1995, 11.8.1995, RDH); TQ3360 (11.8.1995,
RDH); **Riddlesdown Quarry** TQ3359 (19.8.1987, GAC), (21.7.1997, 8.8.1997, RDH);
Caterham (6.9.1936, LP); **Caterham, Happy Valley** TQ3456 (1.9.1985, RKAM);
Warlingham TQ3458 (23.7.1995, RKAM); **Sanderstead Plantation** TQ3461
(24.8.1985, RKAM); **Croham Hurst** TQ3462 (13.8.1985, 21.8.1985, 9.8.1986,
24.8.1986, 27.8.1988, GAC); near **Godstone** TQ3553 (30.8.1993, RKAM); **Nunhead
Cemetery** TQ3575 (9.7.1994, RDH); **Selsdon Wood** TQ3661 (29.8.1985, GAC); **South
Hawke** TQ3753 (29.8.1994, RKAM); **Tatsfield** TQ4256 (3.9.1988, RKAM).

Cheilosia variabilis (Panzer, 1798)

Number of records: 248
Surrey Status: Common
Flight times: April – August
Peak: May

This is a common woodland hoverfly whose
larvae mine the stems of figwort plants. Adults
are common in May and June and are
frequently found sunning themselves on
foliage; the long wings and long, narrow
proportions of the adults make them quite
distinctive, but field identifications should be
confirmed by microscopic examination until one is thoroughly familiar with the genus.
Although widely distributed, *C. variabilis* is absent from much of the London suburbs,
reflecting the range of the two figwort species with which it is associated (see Burton,
1983). There is a single record of this species as a prey item of *Empis tessellata* (Diptera,
Empididae) (Hobby & Smith, 1961).

Flower vists: creeping buttercup, wood spurge (LP, 1957), cow parsley, hogweed, common
figwort, water figwort

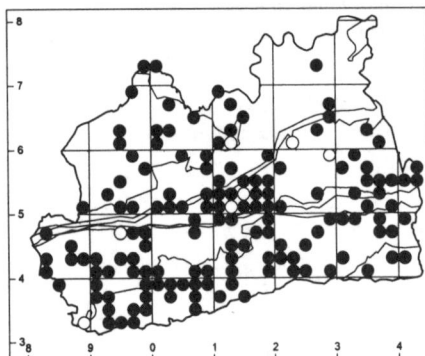

Cheilosia velutina Loew, 1840

Nationally Scarce
Number of records: 11
Surrey Status: Rare
Flight times: July – August
Peak: July

This is a very difficult hoverfly to identify and one which may have been under-recorded. When specimens of *C. velutina* are taken, vouchers should be retained. The main association of adults of this hoverfly appears to be with the same umbellifer flowers that are visited by *C. proxima*.

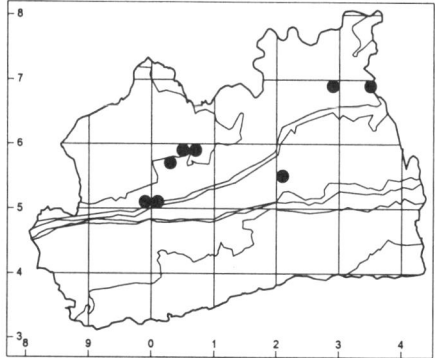

RECORDS: **Guildford, River Wey** SU9951 (3.8.1995, RKAM); near **Guildford** TQ0051 (22.7.1989, GAC/RKAM); **Cartbridge** TQ0256 (22.7.1989, PLTB); **Send** TQ0356 (1.8.1987, 14.8.1988, GAC); **Wisley** TQ0558 (17.7.1989, GAC); **Wisley RHS Gardens** TQ0658 (1982, AJH); **Walton on the Hill** TQ2155 (6.8.1989, RKAM); **Mitcham Common** TQ2868 (23.7.1988, RKAM); **South Norwood Country Park** TQ3568 (2.7.1988, GAC).

Cheilosia vernalis (Fallén, 1817)

Number of records: 207
Surrey Status: Common
Flight times: March – September
Peaks: April, <u>August</u>

This common hoverfly is most frequently found at the flowers of yarrow in the summer and lesser celandine in the spring.

Flower visits: buttercup spp., lesser celandine, hawthorn (LP, 1950a), fool's watercress, wild parsnip, common fleabane, yarrow

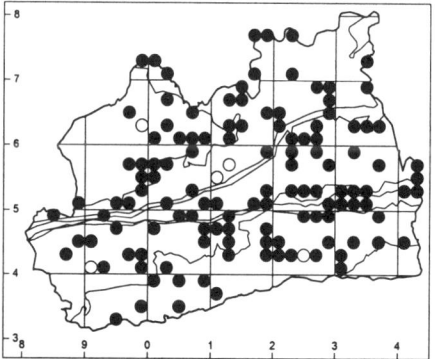

[*Cheilosia vicina* (Zetterstedt, 1849) = *Cheilosia nasutula* Becker, 1894

There is a single published record from Great Bookham Common (LP, 1950a), but this species is mainly northern and the record is therefore extremely dubious and has been discounted.]

Cheilosia vulpina (Meigen, 1822)

Number of records: 55
Surrey Status: Local
Flight times: May – September
Peaks: May, August

Female *C. vulpina* are distinctive in the field on account of their size and the well-developed dust bars on the first two tergites, but males are more easily overlooked. All specimens should be confirmed by more critical examination under the microscope. This hoverfly is most frequently found at hogweed and seems to be most common in dry locations in scrubby or wooded localities.

Flower visits: cow parsley, fool's water-cress, wild parsnip, hogweed, wild carrot

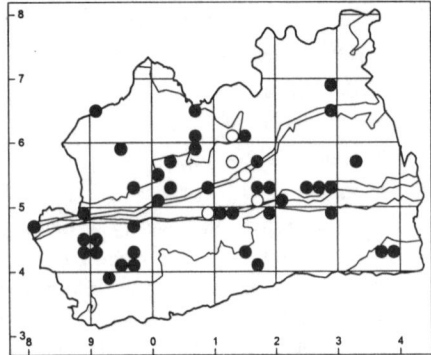

Genus *Ferdinandea*

There are only two species, one of which (*F. cuprea*) is generally common and distinctive with a bronzy abdomen and vaguely striped thorax. The larvae of both are associated with sap runs, at which adults may also be found.

Ferdinandea cuprea (Scopoli, 1763)

Number of records: 160
Surrey Status: Common
Flight times: April – September
Peaks: June, August

This is a woodland hoverfly which occurs most frequently in the woods of the Chalk and the Low Weald, but can be found in many other wooded and scrubby localities. Adults visit a range of yellow composites and umbellifers; they can also be found sunning themselves on tree trunks and at sap runs.

Flower visits: buttercup spp., greater stitchwort, perforate St John's-wort, field bindweed, hedge bindweed, creeping thistle, nipplewort, hawkweed oxtongue, perennial sow-thistle, dandelion, smooth hawk's-beard

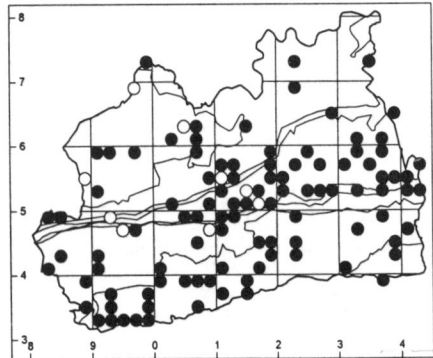

Ferdinandea ruficornis (Fabricius, 1775)

Nationally Scarce

Number of records: 5

Surrey Status: Rare

Flight times: April – August

Our knowledge of this hoverfly in Surrey is confined to just a few records. It is very clearly a visitor to sap runs, with two records from an oak infested by goat moth at Wisley RHS gardens (AJH), and a record of two females "on a tree oozing sap" on Roehampton Common (AWJ). The record from Box Hill (KNAA) is not supported by a specimen, but has been accepted on the basis of the recorder's experience. This is not a regularly recorded species on a national scale and may be overlooked; it is smaller and darker than *F. cuprea* with which it may be confused, so small specimens of the latter should be retained for critical examination.

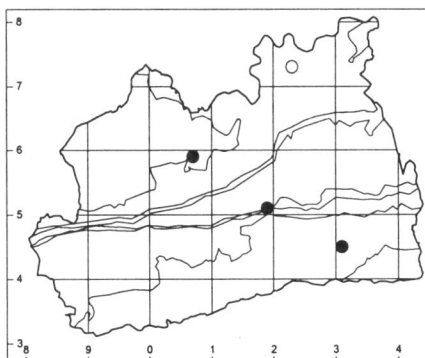

Conservation: In keeping with that for goat moth, infested trees should be retained, although if seriously affected, they may have to be made safe by pollarding. Other trees with sap runs are also important for this and other species of sap-dwellers, and should be retained wherever possible because sap runs are not necessarily indicative of a severely weakened tree. Only in extreme circumstances, where public safety is an issue, should felling outright be considered as a last resort.

RECORDS: **Wisley RHS Gardens** TQ0658 (16.8.1988, 17.8.1988, AJH); **Box Hill** TQ1851 (22.7.1997, KNAA); **Roehampton Common** TQ2273 (26.6.1949, AWJ); **Outwood** TQ3144 (21.4.1994, RDH).

Genus *Portevinia*

Portevinia maculata (Fallén, 1817)   PLATE 2

Number of records: 34
Surrey Status: Scarce
Flight times: May – June
Peak: May

The deep stream gorges of the Low Weald are the principal habitat for this hoverfly which is closely associated with the bulbs of ramsons. *P. maculata* was formerly thought to be very local in England and Wales, but in Surrey it can be found at almost any locality where ramsons occurs, as is demonstrated by comparing the maps of the fly and its host plant. Ramsons is usually found on calcareous soils and is consequently absent from much of the county north of the Downs, hence the absence of *P. maculata* from this area. I have found that it helps to leave the car window down so that one can detect the ramsons by smell whilst driving past likely sites, but in recent years have also used records of the plant held by SWT to track down *P. maculata*.

Flower visits: ramsons

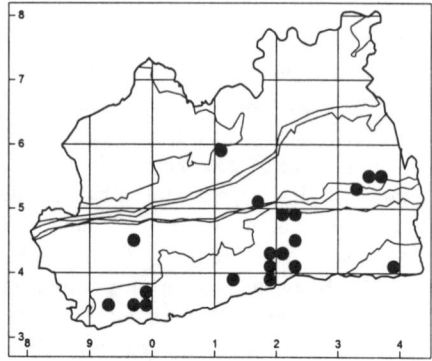

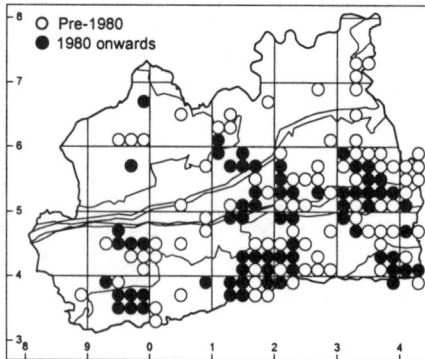

The distribution of ramsons in Surrey
based on data held by SWT and atlases
of the flora of Surrey
(Lousley, 1976; Burton, 1983).

Genus *Rhingia*

There are two species which can be relatively easily separated, once known. They are rather globular reddish flies with an extended mouth edge which gives them a distinctive appearance. These are the so-called "Heineken flies" of Whiteley (1987).

Rhingia campestris Meigen, 1822

Number of records: 626
Surrey Status: Ubiquitous
Flight times: April – September
Peaks: <u>May</u>, August

This is a common species whose larval stages are associated with cow dung, which explains the lack of records from much of the London area. The few records from the London area suggest that it may be a mobile species or that other animal dung may also be used. It is a frequent visitor to flowers such as water mint and can occur in phenomenal numbers. There are two distinct broods, in May/June and in August, and in many years there is a very marked absence of this species in July. In 1996 the summer brood apparently failed almost entirely, which may have been a reflection of the extremely hot and dry conditions. This would not appear to be wholly unusual, as there are various references to a similar dearth of records in 1948, both in Surrey and elsewhere (Laurence, 1948; LP, 1948). Analysis of the frequency of this species, based simply on the proportion of the total number of records for Surrey for each year over the study period (1985-1997), suggests that this species' occurrence may also be cyclic (see *Figure 13*). But, an analysis of the national data-set linked to overall levels of recording for each year does not support this hypothesis (S. Ball, *pers. comm.*). There would also seem to be an overall drop in the numbers of records for this species since 1985, which may be due to recent droughts or could also be associated with the impact of avermectins (chemical treatment for insect parasites in cattle) on the dung-feeding fauna.

Flower visits: traveller's-joy, lesser celandine (LP, 1950a), water chickweed, ragged robin, violet sp. (LP, 1950a), hedge mustard (LP, 1950a), garlic mustard, cuckoo-flower, St Dabeoc's heath, bilberry, primrose (LP, 1950a), bramble, blackthorn (LP, 1950a), hawthorn, great willowherb, wood spurge (LP, 1957), sycamore, herb robert, Indian balsam, cow parsley, wild angelica (Uffen, 1969), field bindweed, hedge bindweed, large bindweed, green alkanet, hedge woundwort, white dead-nettle, common hemp-nettle, bugle, ground-ivy, self-heal (LP, 1950a), wild marjoram, water mint (LP, 1950a), garden lavender, butterfly-bush, red bartsia, devil's-bit scabious, welted thistle, spear thistle, common knapweed, dandelion, ragwort spp., bluebell (LP, 1950a)

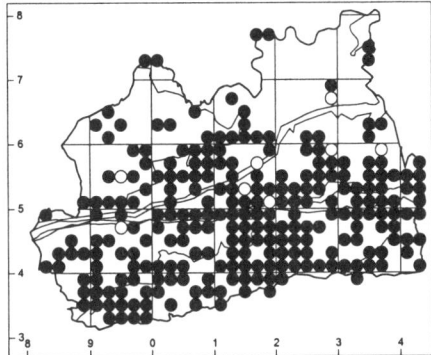

85	86	87	88	89	90	91	92	93	94	95	96	97
60	71	108	73	38	7	7	33	54	65	25	9	11
2.8%	3.7%	2.7%	2.9%	1.4%	0.7%	0.67%	2.3%	2.8%	2%	1.2%	1.2%	1.7%

Figure 13. Annual counts of records of *Rhingia campestris* from Surrey and the percentage of all records for Surrey in each year from 1985 to 1997

Rhingia rostrata (Linnaeus, 1758)

Red Data Book 3
Number of records: 18
Surrey Status: Rare
Flight times: May – September
Peaks: May, <u>August</u>

This species is superficially similar to *R. campestris* and may be overlooked as such. The larval habits are unknown, but carrion has been suggested as a possibility (Stubbs and Falk, 1983); this seems unlikely because adults exhibit two distinct peaks at times when carrion is unlikely to be plentiful. The records to date clearly indicate that this is a woodland species, so an association with woodland mammal dung seems a plausible suggestion, the most obvious being with badger latrines. The first records for Surrey were from Chelsham Wood when Coe (1939b) took a total of nine specimens over three visits between 11.6 and 15.8.1939, and on a variety of other dates until 1953 (Coe, 1961). Most subsequent records are of single individuals, but there are reports of massed occurrence as reported by Coe (1961) who took 15 males and 14 females on 14.5.1961 at Selsdon Wood. During this survey one such incident of massed occurrence was observed at South Hawke, TQ3753, on 5.9.1992 when this was the commonest species in a woodland clearing (RKAM/GAC).

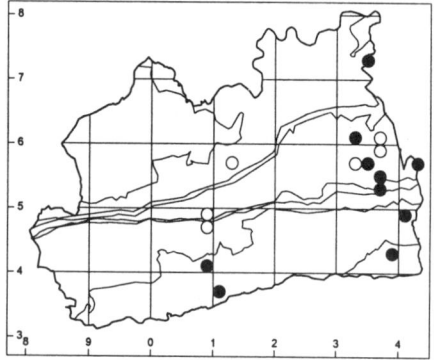

Flower visits: hawthorn (Coe, 1961), dogwood (Coe, 1961), butterfly-bush, nettle-leaved bellflower (LP, 1956a), devil's-bit scabious (LP, 1950a)

RECORDS: **Ewhurst Woods** TQ0840 (5.5.1995, RKAM); **Gomshall** TQ0847 (no date, PJC); **Hackhurst Downs** TQ0948 (2.7.1966, SBRC); **Somersbury Wood** TQ1037 (7.6.1987, GAC); **Great Bookham Common** (29.9.1946, LP); **Coulsdon Common** (18.8.1956, LP); **Riddlesdown** TQ3260 (13.6.1996, RDH); **Warlingham** TQ3557 (29.8.1994, RKAM); **Chelsham Wood** (11.6.1939, 13.8.1939, 15.8.1939, [Coe, 1939b], 23.5.1940, 19.8.1951, 24.5.1953 [Coe, 1961]); **Sydenham Hill Woods** TQ3472 (16.6.1988, AG); **Great Church Wood** TQ3654 (20.5.1987, GAC); **Selsdon Wood** TQ3661 (14.5.1961, [Coe, 1961]); **South Hawke** TQ3753 (5.9.1992, RKAM/GAC); **Mill Wood, Lingfield** TQ3942 (30.5.1994, RDH); **Staffhurst Wood** TQ4148 (30.8.1987, RKAM); **Tatsfield** TQ4256 (30.8.1987, GAC).

Chrysogastrini

Many species occupy wet habitats and larval associations include sap runs, decaying vegetable matter, dead wood and damp soil around wetlands. Many, especially *Chrysogaster*, *Melanogaster*, *Neoascia*, *Orthonevra*, *Lejogaster* and *Riponnensia*, are wetland associates and can occur in large numbers.

Genus *Brachyopa*

This is a group of small and generally poorly recorded species with orange-brown abdomens whose larvae are associated with sap runs on a variety of trees and are readily identified using the key by Rotheray (1996).

Brachyopa bicolor (Fallén, 1817)

Red Data Book 3
Number of records: 2
Surrey Status: Rare
Flight times: May

This is a scarce species which occurs sparingly in southern England. It was thought to be associated with old woodland, but single records from Kew Gardens and Sheerwater suggests that this is not so. The Kew record appears to be the first post-1970 record for *B. bicolor* in the London area (Plant, 1986).

RECORDS: **Basingstoke Canal, Sheerwater** TQ0260 (1982, AJH); **Kew Gardens** TQ1876 (9.5.1989, RBH).

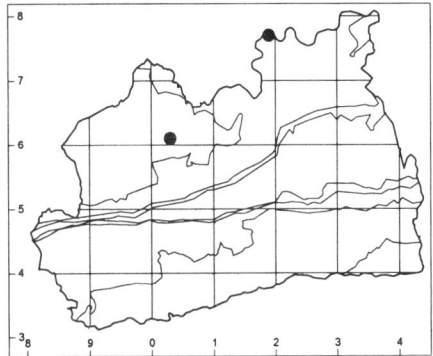

Brachyopa insensilis Collin, 1939   PLATE 7

Nationally Scarce
Number of records: 39
Surrey Status: Local
Flight times: May – June
Peak: May

At one time *B. insensilis* was closely associated with sap runs on elm and its demise could have followed that of the English elms to Dutch elm disease. But recent investigations of crusty sap runs on horse chestnut have yielded large numbers of *B. insensilis* as adults in May and June, and larvae during the winter.

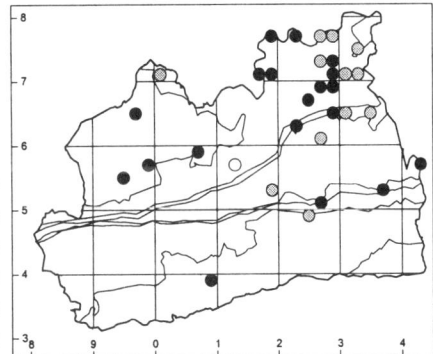

I have observed a female at Tooting Bec Common on 24.5.1992 ovipositing on dryish bark adjacent to a sap run, and not directly into the sap run. There is also a single record of this species being associated with oak (JSD), which may help to explain the occurrence of this species near Ewhurst Green where no horse chestnut was present in the immediate area. Now that the methods of finding the larvae have been identified, its national status clearly needs to be revised. It appears to occur widely on suitable trees in both suburban and rural locations, although not all sap runs will support it. The highest concentration of records is from the London area where horse chestnut is commonly planted in parks and streets. I have found that the red-flowered variety of horse chestnut is far less prone to sap runs and do not recall finding a decent sap run on one of these on any occasion. Many records are of larvae, but on no occasion have I successfully reared them; some records must therefore be considered with caution. The map for this species consequently includes a further class, with grey circles representing larval records that require confirmation. In commenting on this text, Graham Rotheray offered the following advice about rearing *Brachyopa* larvae: "they are long-lived, perhaps requiring two years to mature; keep them in sap-run material in open, not sealed, airtight vessels, in cool dark conditions, and change the sap about every six weeks (and) expect to have to keep them for twelve months or more".

Conservation: This species' close association with sap runs on horse chestnut illustrates the importance of retaining trees which show signs of damage. Many sap runs are temporary and probably do little damage to the tree, so this alone should not be used as an indicator of a diseased and dangerous tree.

RECORDS: **ADULTS: Brookwood Common** SU9555 (2.6.1996, JSD); **Chobham Place Wood** SU9664 (31.5.1992, AJH); **Woking** SU9957 (13.6.1981, AJH); **Wisley RHS Gardens** TQ0659 (1980-1981, 11.6.1996, AJH); **Wisley Common** TQ0658 (17.6.1971, AES); near **Ewhurst Green** TQ0938 (7.6.1987, GAC); **Great Bookham Common** (26.6.1966, LP); near **Teddington** TQ1671 (23.6.1996, RKAM); **Richmond Park** TQ1871 (6.6.1992, RKAM); **Kew Gardens** TQ1876 (1970, AES), (16.5.1989, RBH); **Nonsuch Park** TQ2263 (31.5.1992, RKAM); **Barnes Common** TQ2276 (30.5.1992, RKAM); **Morden Park** TQ2467 (21.5.1989, RKAM); **Morden Hall Park** TQ2668 (24.5.1992, RKAM); **Gatton** TQ2751 (25.5.1992, GAC).; **Beddington Park** TQ2865 (24.5.1992, 29.5.1995, RKAM); **Mitcham** TQ2868 (24.5.1992, RKAM); **Tooting Bec Common** TQ2871 (24.5.1992, RKAM), TQ2972 (24.5.1992, RKAM); near **Godstone** TQ3652 (1993, CWP); **Tatsfield** TQ4256 (25.5.1992, RKAM/GAC). **LARVAE:** near **Runnymede** TQ0070 (14.3.1993, RKAM); **Cockshot Wood** TQ1853 (22.6.1996, RKAM); **Reigate** TQ2549 (20.3.1993, RKAM); **Woodmansterne** TQ2760 (20.3.1993, RKAM); **Wandsworth Common** TQ2773 (30.3.1997, RKAM); **Battersea Park** TQ2777 (21.3.1993, RKAM), TQ2877 (21.3.1993, RKAM); **Streatham Common** TQ3070 (21.3.1993, 16.4.1995, RKAM); **Wandle Park, Croydon** TQ3165 (9.6.1996, RKAM); **Sunray Park** TQ3274 (16.4.1995, RKAM); **Gypsy Hill Park** TQ3371 (16.4.1995, RKAM); **Lloyd Park** TQ3464 (21.6.1996, RKAM).

Brachyopa pilosa Collin, 1939

Nationally Scarce
Number of records: 15
Surrey Status: Rare
Flight times: April – June
Peak: May/June

To find this fly one must normally examine recently fallen beech trunks whose bark remains firmly attached to the timber, and also sap runs on living and recently cut stumps; I once found this species in exceptional numbers around cut beech stumps with large sap runs at Oaken Grove (TQ1149) on 15.6.1987. There is one record of a puparium found at Ranmore "under a thin outer flake of bark surface" which produced an adult *B. pilosa* (McLean & Stubbs, 1990). In such locations *B. pilosa* can be remarkably common. Although it is mainly confined to woodlands on the Chalk, *B. pilosa* does occur elsewhere; for example at Merrow Common where it appeared to be associated with white/grey poplar, *Populus alba/canescens*, and at Hindhead Common where it was associated with cut and stacked birch, *Betula* spp.

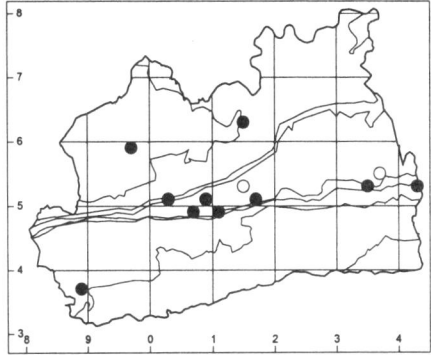

Conservation: It is essential to retain fallen beech timber *in-situ*, and also not to remove trees with obvious sap runs. Fallen beeches are usually found in open clearings and are highly susceptible to desiccation, but this does not appear to be a problem for *B. pilosa* which seems to be an opportunistic species which colonises suitable timber. It may be possible to extend the life of suitable timber by moving larger boughs into more shady locations.

RECORDS: **Hindhead Common** SU8936 (28.5.1988, GAC); **Barrs Lane, Knaphill** SU9659 (1982, AJH); **Merrow Common** TQ0251 (28.4.1990, RKAM/PLTB); **Silent Pool** TQ0648 (16.5.1992, RKAM); **The Sheepleas** TQ0851 (15.5.1994, AJH); **Oaken Grove** TQ1149 (15.6.1987, RKAM); **Ranmore** (no date, IFGM/AES); **Norbury Park** TQ1553 (1949, JHPS); **Great Oaks** TQ1562 (13.4.1997, GAC); **Burford Bridge** TQ1751 (1.6.1987, GAC); **Pilgrim Fort** TQ3453 (12.6.1987, GAC); **South Hawke** TQ3754 (31.5.1971, AES); **Limpsfield Chart** TQ4252 (1.6.1986, GAC), (21.6.1989, RKAM).

Brachyopa scutellaris Robineau-Desvoidy, 1843

Number of records: 39
Surrey Status: Local
Flight times: April – June
Peak: May

This widespread woodland species is mainly found on the Chalk and the Low Weald. It can occasionally be very common as at Wanborough Wood, SU9149 (29.4.90, RKAM), where it was seen in numbers flying around birch trunks, and at Ewhurst Woods, TQ0840 (RKAM), where it was the commonest hoverfly on 5.5.1995 flying around oak trunks. In addition to obvious sap runs, *B. scutellaris* can be found on sunlit sycamore leaves and in dappled sunlight around trunks of beech and oak with no obvious sap run. This hoverfly may have a very restricted emergence period as it has been found to occur widely on certain days and yet to be absent in apparently suitable habitat only a few days later. Specimens taken in 1995 differed from those recorded in previous years in that the humeri are dark and not yellow; the antennal pits are the same however. Unlike other *Brachyopa* species, the larvae seem to prefer sap runs close to the ground (Rotheray, 1996).

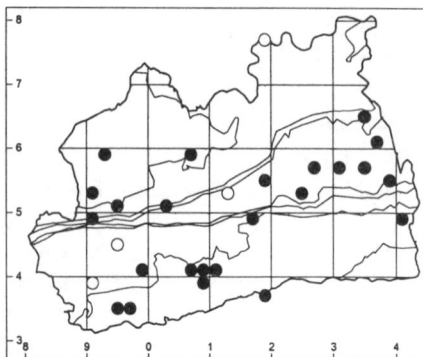

Genus *Chrysogaster*

Formerly this genus included the two *Melanogaster* species (*M. aerosa* and *M. hirtella*). They are small dark hoverflies with a slight metallic sheen on the thorax and edges of the tergites, which often occur together with *Cheilosia* spp and may be mistaken for the latter by inexperienced recorders until the facial features are recognised. The larvae are aquatic, but given the widespread distribution of such species as *C. solstitialis* they must be associated with small water bodies or muddy soils.

Chrysogaster cemiteriorum (Linnaeus, 1758) = *chalybeata* Meigen, 1822

Number of records: 47

Surrey Status: Local

Flight times: June – August

Peak: July

This widely distributed hoverfly is commonest in damp locations, especially woodland. Current records suggest a clumping of distribution, mainly concentrated south of the North Downs and on the south-west Surrey heaths, but absent from most of the Downs themselves.

Flower visits: burnet-saxifrage, wild parsnip, hogweed, upright hedge-parsley

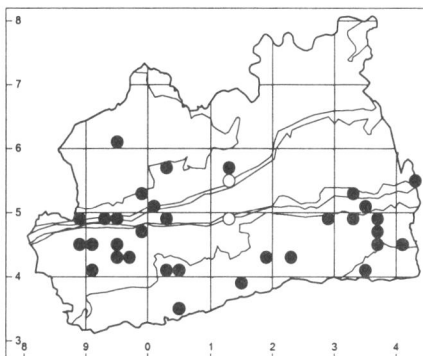

Chrysogaster solstitialis (Fallén, 1817)

Number of records: 321

Surrey Status: Common

Flight times: May – September

Peak: August

This is a widespread woodland species which frequently occurs in numbers at umbellifers, especially hemlock water-dropwort, wild angelica and hogweed. Its apparent absence from much of the London suburbs reflects the overall lack of woodland as shown in the map of woodland distribution in the introduction.

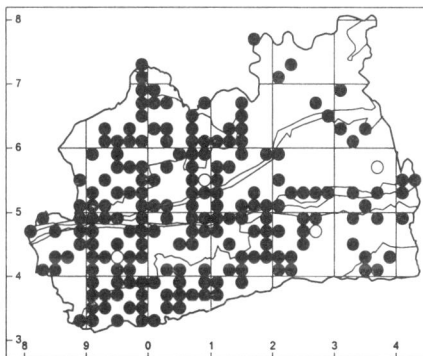

Flower visits: cow parsley, burnet-saxifrage, ground-elder, hemlock water-dropwort, fool's water-cress, wild angelica, hogweed, upright hedge-parsley, wild carrot, yarrow, ragwort spp.

Chrysogaster virescens Loew, 1854

Number of records: 12

Surrey Status: Rare

Flight times: May – August

Peak: June

This heathland hoverfly appears to be quite widely distributed. It can often be found at yellow flowers including those of broom. Most records are for June, but there is also a published record for late August (Fry & Denton, 1992).

Flower visits: broom

RECORDS: **Tilford Reeds** SU8643 (28.5.1988, RKAM); **Hankley Common** SU8841 (9.6.1991, JRD); **Thursley Common** SU9041 (14.6.1969, PJC), (14.8.1989, JSD); **Frith Hill** SU9058 (22.5.1988, RKAM/GAC); **Bagshot** (8.6.1934, OWR), SU9063 (29.5.1994, RKAM); **Brentmoor Heath** SU9261 (1997, JSD); **Whitmoor Common** SU9853 (23.6.1996, AJH); **Esher Common** TQ1262 (16.6.1988, RKAM); **Oxshott** (24.6.1951, LP).

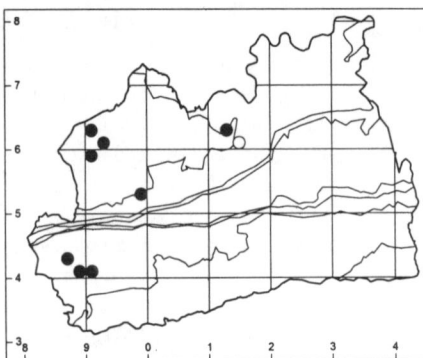

Genus *Lejogaster*

These are small, distinctively metallic species, both of which are associated with wetlands. Neither is common, at least in part reflecting the paucity of wet grasslands in Surrey.

Lejogaster metallina (Fabricius, 1781)

Number of records: 24

Surrey Status: Scarce

Flight times: May – September

Peak: May

Although *L. metallina* is widely distributed in the UK, it is clearly very uncommon in Surrey. It is usually found in wet locations but has also been found in arable situations at scentless mayweed flowers.

Flower visits: buttercup spp., bastard cabbage, scentless mayweed

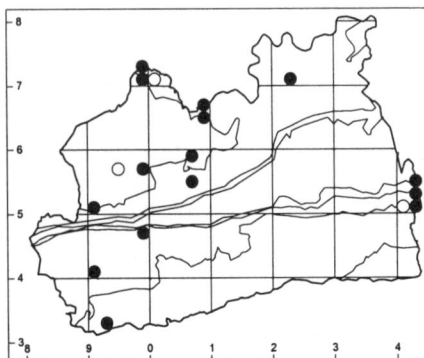

Lejogaster tarsata (Megerle *in* Meigen, 1822)
= *Lejogaster splendida* (Meigen, 1822)

Nationally Scarce

Number of records: 4

Surrey Status: Rare

Flight times: June – July

There are a number of old records for this species at Runnymede which appears to be the only known locality in Surrey. This is essentially a coastal species with scattered inland records, and it is surprising that it occurs so far inland where tidal and saline influences are weak. Further recording along the Thames may reveal new localities and the possibility of this species' occurrence should always be borne in mind when recording along the Thames.

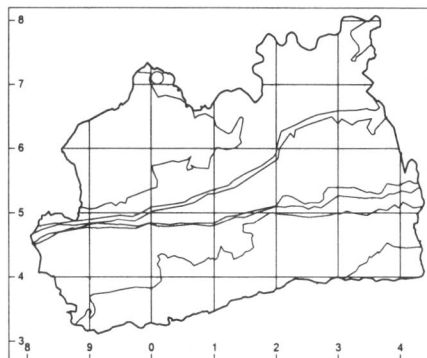

RECORDS: **Runnymede** TQ0071 (18.6.1974, 13.7.1973, 19.7.1973, PJC), (23.6.1974, COH).

Genus *Melanogaster*

These are small black flies, not dissimilar to *Cheilosia*, but distinguished by the metallic edges to tergites 2 to 5 and the flattened dull abdomen. The larvae are associated with damp or wet habitats.

Melanogaster aerosa (Loew, 1843) = *Chrysogaster macquarti*: auctt. misident.

Nationally Scarce

Number of records: 9

Surrey Status: Rare

Flight times: July – September

Peak: July

There are very few records of this heathland hoverfly which may be under-recorded owing to its late flight period at a time when the heaths get very hot and difficult to work and hence are not regularly visited by dipterists.

RECORDS: **Thursley Common** SU9041 (27.7.1966, 30.8.1969, 06.9.1970, PJC), (1966 - 1970, AES), (14.8.1990, JSD), (1992, RF); **Pirbright Ranges** SU9260 (15.8.1989 SJF); **Chobham Common** SU9665 (16.8.1969, PJC); **Chobham Longcross** SU9765 (1.8.1987, GAC).

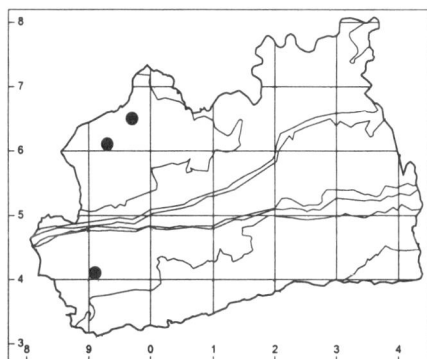

Melanogaster hirtella (Loew, 1843) = *Chrysogaster hirtella* (Loew, 1843)

Number of records: 153
Surrey Status: Common
Flight times: May – August
Peak: July

This common hoverfly is mainly found in marshy situations at buttercup flowers. Its larvae have been found under plants of marsh marigold (G. Rotheray, *pers. comm.*), but it must have a wider range of associations given its widespread distribution. It is probably under-recorded because it has a short flight period, but is seemingly commoner in west Surrey and perhaps on the wealden clays. Male *M. hirtella* have a slight "nose" and I have found that individuals of this species are often keyed out as *Cheilosia antiqua* by the inexperienced recorder.

Flower visits: meadow buttercup, creeping buttercup, buttercup spp.

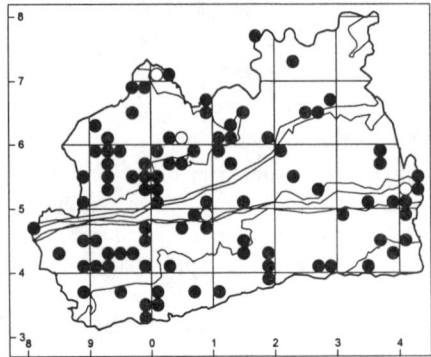

Genus *Myolepta*

Myolepta dubia (Fabricius, 1805) = *Myolepta luteola* (Gmelin, 1790)

Nationally Scarce
Number of records: 13
Surrey Status: Rare
Flight times: June – August
Peak: August

In my experience, *M. dubia* has a very short flight period and there are grounds for believing that it is in fact a good deal more widespread than the records suggest. Adults are usually found at umbel flowers and may be overlooked as *Cheilosia impressa* or possibly a late *C. albitarsis* because the yellow tergites are masked by the wings which assume the conspicuous yellowish wing bases typical of *C. impressa*. The larvae inhabit wet rot-holes in beech, horse-chestnut and oak (Rotheray, 1993) and are easy to identify. There are two breeding records from Surrey from Putney Heath, both of larvae "from a rotting tree, probably elm" (A.W. Jones, *pers. comm.*). Recording larvae may in fact prove more worthwhile as this is one of a suite of species which occur more commonly as larvae than as adults.

Conservation: Avoiding popular arboricultural practice in parkland situations would be a positive advantage for species such as *M. dubia* which breed in rot-holes, but in practice

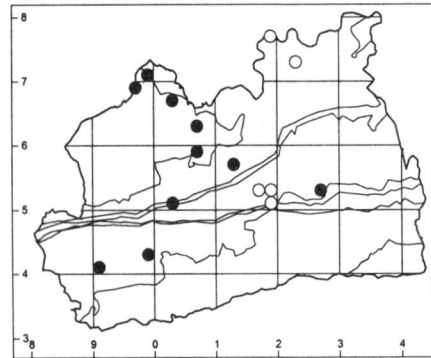

public safety is an important issue. Rot-holes should be left to develop naturally and not filled with concrete as is often the case.

Flower visits: hogweed, upright hedge-parsley, thistle spp. (Withers, 1983)

RECORDS: **Thursley** SU9040 (1990, RF); **Windsor Great Park** SU9768 (1.8.1987, SRM); **Busbridge** SU9842 (18.6.1994, RKAM); near **Englefield Green** SU9971 (1.8.1987, GAC); **Merrow Common** TQ0251 (5.7.1987, RKAM); **St. Anne's Hill** TQ0267 (1.8.1987, RKAM/GAC); **Wisley RHS Gardens** TQ0659 (9.7.1997, AJH); **Weybridge** TQ0663 (2.8.1997, RDH); **Great Bookham Common** TQ1256 (22.6.1988, GAC); **Mickleham** (no data, Withers, 1983); **Box Hill** TQ1851 (29.8.1977, AES); **Kew** (26.6.1961, MS);**Putney Heath** TQ2273 (larvae 1.6.1949, 5.6.51, AWJ); **Upper Gatton** TQ2753 (2.8.1987, RKAM).

Genus *Neoascia*   PLATE 12

These are tiny black or black-and-yellow species, some of which have slight shading on the wings. Their larvae are associated with decaying vegetation, mainly in wetlands. *Neoascia tenur* and *N. meticulosa* are seemingly concentrated in west Surrey, but it is noteworthy that just 11 tetrads (21%) coincide. This suggests that the two species have quite differing requirements. I suspect that *N. tenur* favours more acid situations whilst *N. meticulosa* may be more closely associated with basic influences or alluvial soils, as is indicated by the high concentration of records in the valley of the River Wey around Guildford and Godalming.

Neoascia podagrica (Fabricius, 1775)

Number of records: 305

Surrey Status: Common

Flight times: April – October

Peaks: <u>May</u>, <u>August</u>

Although widespread and often common, *N. podagrica* seems to be more frequent on the wealden clays and in the damper parts of west Surrey. It is seemingly scarce in north-east Surrey for reasons that are not immediately apparent. The larvae are possibly associated with a wide range of damp decaying plant material and this may help to explain the apparent preponderance of records on the Low Weald where conditions are wetter.

Flower visits: wood anemone, buttercup spp., garlic mustard, water-cress, bramble, cow parsley, fool's water-cress, wild parsnip, hogweed, upright hedge-parsley, cat's-ear, scentless mayweed

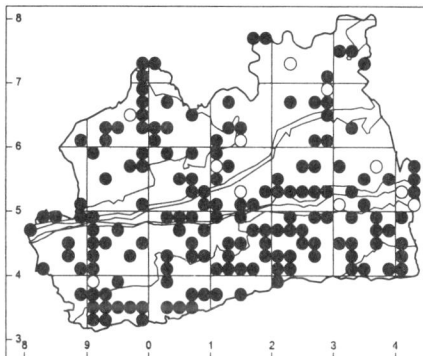

Neoascia tenur (Harris, 1780)

Number of records: 69
Surrey Status: Local
Flight times: April – August
Peaks: June, August

This is a common species in west Surrey which appears to be most abundant on wetland on the heaths and in acidic habitats. It is seemingly scarce on the clay of the Low Weald and much of the Chalk and London Clay.

Flower visits: creeping buttercup (LP, 1950a), tormentil, hogweed

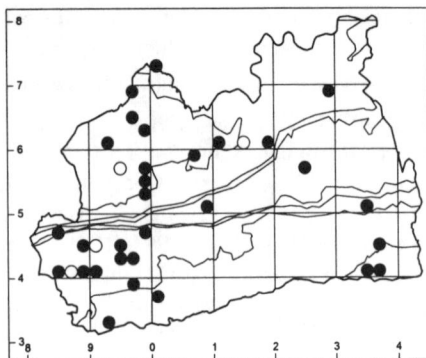

Neoascia geniculata (Meigen, 1822)

Nationally Scarce
Number of records: 11
Surrey Status: Rare
Flight times: May – August
Peaks: May, August

This is a very difficult species to identify and one which may be under-recorded for this reason. It is a wetland species which can occasionally be found in large numbers, for example at Blindley Heath pond.

Flower visits: creeping buttercup (LP, 1950a)

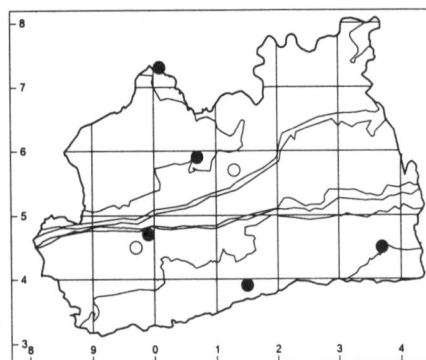

RECORDS: **Farncombe** (3.8.1923, AT); **Shalford** SU9947 (15.7.1989, RKAM); **Runnymede** TQ0072 (10.8.1986, DJG); **Wisley Common** TQ0658 (15.6.1986, RKAM); **Great Bookham Common** (12.8.1951, SW), (25.5.1941, LP); **Vann Lake** TQ1539 (9.7.1988, RKAM); **Blindley Heath Pond** TQ3645 (12.7.1986, 10.8.1986, RKAM/GAC).

Neoascia interrupta (Meigen, 1822)

Nationally Scarce
Number of records: 10
Surrey Status: Rare
Flight times: April – August
Peaks: <u>June</u>, August

Neoascia interrupta is a relatively recent addition to the British list and one of the more obvious members of the genus. Although widely distributed, it is probably genuinely scarce. Personal observations suggest that it is closely associated with beds of bulrush, *Typha* spp.

RECORDS: **Farnham Park** SU8348 (29.5. - 5.7.1997, JSD); **Ash Wharf** SU8851 (2.5.1995, GM), SU8951 (17.6.1996, GAC)**; Runnymede** TQ0072 (10.8.1986, DJG); **Wisley RHS Gardens** TQ0659 (23.4.1997, AJH)**; Great Bookham Common** TQ1256 (17.6.1987, GAC)**; Epsom Common** TQ1860 (29.6.1986, RKAM); **Bay Pond** TQ3551 (6.8.1986, 10.6.1987, 26.6.1988, GAC).

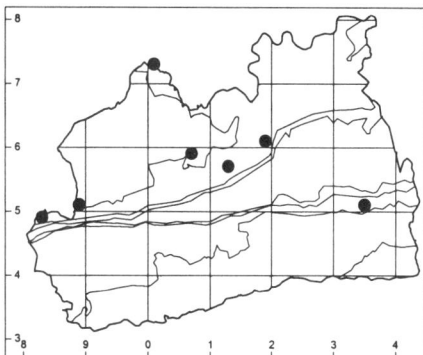

Neoascia meticulosa (Scopoli, 1763)

Number of records: 56
Surrey Status: Local
Flight times: April – August
Peak: May

This is a widespread wetland hoverfly which is easily confused with *N. tenur*. On occasions it can be very common flying amongst low vegetation. Most records are from west Surrey with the main clump around the valley of the River Wey where wet pasture remains widespread.

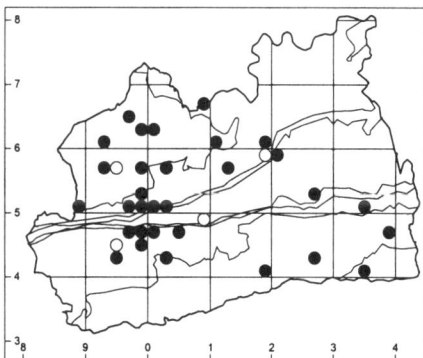

Neoascia obliqua Coe, 1940

Nationally Scarce
Number of records: 3
Surrey Status: Rare
Flight times: April – June

Until recently, the ecology of this species was poorly understood, but it has now been suggested that it is associated with butterbur, *Petasites hybridus* (Stubbs, 1996). As a consequence, it has proved to be more widespread than previously thought. The first

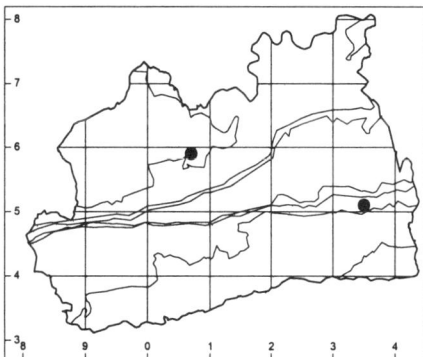

Surrey record was from a small alder carr near Godstone in 1996; since then it has been taken from an old oxbow at Wisley. Unfortunately, this discovery came too late to allow a full survey of suitable sites, and it is very likely that this fly will prove to be more widespread given suitable efforts to find it. Retention of specimens of *N. podagrica* is essential when recording from sites where butterbur occurs; the distinction between this and *N. obliqua* is clear when examining the chitinous bridge behind the hind coxae.

Flower visits: buttercup spp.

RECORDS: **Wisley RHS Gardens** TQ0659 (23.4.1997, AJH); near **Godstone** TQ3550 (9.6.1996, RKAM), (17.6.1996, GAC).

Genus *Orthonevra*

These are small but distinctive dark flies with a metallic sheen. The larvae are wetland associates.

Orthonevra brevicornis Loew, 1843

Nationally Scarce
Number of records: 5
Surrey Status: Rare
Flight times: May – June
Peak: June

This wetland hoverfly is most frequently found flying low amongst vegetation, but has also been found at umbel flowers. Records suggest that this species is most likely to be encountered at peaty locations such as marsh adjacent to alder carr, and in fen-like situations. (The rich fen at Tuesley is one of the undiscovered gems in Surrey with seepages, wet rotting timber and great horsetail, *Equisetum telmateia.*)

Conservation: This is one of a suite of species associated with rich fens which are scarce in Surrey. These sites should be identified and safeguarded wherever possible; some may be affected by water abstraction and recent droughts.

Flower visits: hemlock water-dropwort

RECORDS: **Thursley Common** SU9040 (2.6.1968, AES); **Tuesley** SU9642 (25.5.1987, GAC); **Gomshall** TQ0847 (8.6.1986, RKAM/GAC); near **Jayes Park** TQ1441 (5.6.1993, RKAM).

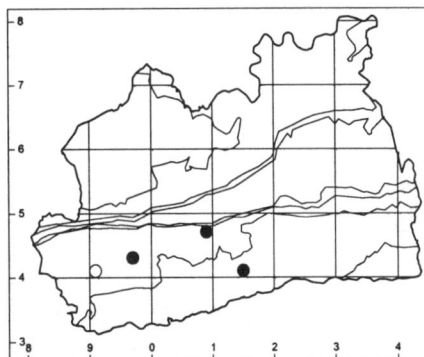

Orthonevra geniculata Meigen, 1830

Nationally Scarce
Number of records: 9
Surrey Status: Rare
Flight times: April – May
Peak: April

This appears to be a species of wet heathland and carr woodland, judging from the small number of records that exist for Surrey. The specimens at Horsell Common were taken at sallow catkins. It is quite distinct and is unlikely to be overlooked, so it is probably genuinely scarce.

Flower visits: sallow

RECORDS: **Chobham Common** SU9665 (27.4.1968, PJC), SU9666 (24.5.1986, AJH); **Egham** (11.5.1947, CHA); **Whitmoor Common** SU9853 (14.5.1995, AJH); **Horsell Common** TQ0160 (15.4.1989, pair *in cop.* PLTB); **Wisley Common** TQ0658 (post-1980, AES); **Esher Common** (no date, OWR); **Oxshott** (22.4.1925, OWR); **Hedgecourt** TQ3540 (27.4.1996, GAC).

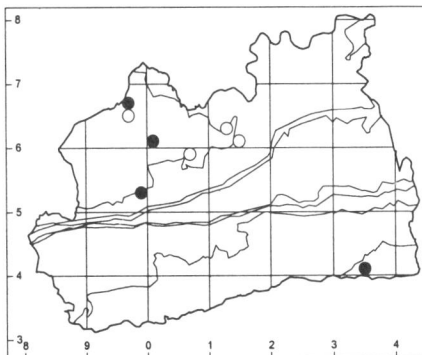

Orthonevra nobilis (Fallén, 1817)

Number of records: 77
Surrey Status: Local
Flight times: May – September
Peak: August

Although one tends to associate the genus with wetter locations, *O. nobilis* is more widely distributed on rough grasslands. It is most common on the Low Weald, but can also be found in extremely dry locations on the scarp of the Chalk. It is most frequently found at umbellifer flowers in midsummer.

Flower visits: buttercup spp., burnet-saxifrage, ground-elder, wild parsnip, hogweed, upright hedge-parsley, wild carrot, yarrow

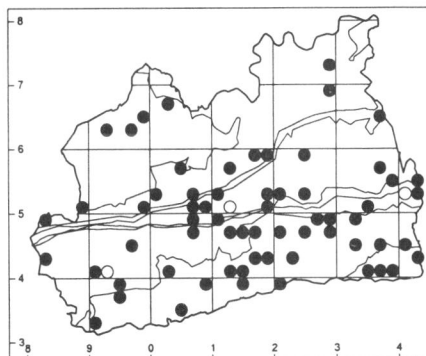

Genus *Riponnensia*

Formerly this species was considered to be part of *Orthonevra*. It is somewhat larger than British members of the latter genus, and is highly distinctive, being rather more strongly metallic coloured. Its larvae are also associated with wet habitats.

Riponnensia splendens (Meigen, 1822) = *Orthonevra splendens* (Meigen, 1822)

Number of records: 77

Surrey Status: Local

Flight times: May – September

Peak: August

This widely distributed hoverfly is often found at the flowers of hogweed, hemlock water-dropwort and wild angelica in wooded localities. *R. splendens* is principally distributed on the Low Weald and London Clay, with few records from the western heaths and the Chalk.

Flower visits: buttercup spp., cow parsley, burnet-saxifrage, ground-elder, hemlock water-dropwort, wild angelica, hogweed, wild carrot

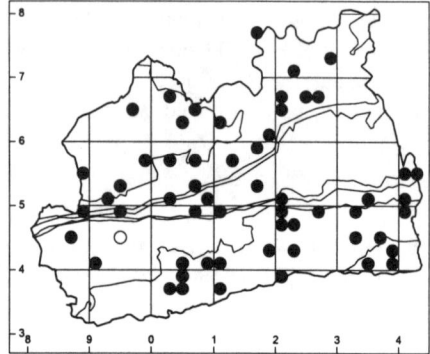

Genus *Sphegina*

There are three species of this group of small narrow-bodied hoverflies which are associated with woodland where they are most commonly found either by sweeping plants such as sanicle, *Sanicula europaea*, or by searching the undersides of other umbels such as wild angelica. They can be found in great numbers on occasions, but more frequently occur sparsely; look for a flower in dappled sunlight and watch for their arrival. There is a fourth member of the genus, *S. sibirica*, recently recognised as British (Stubbs, 1994), and is associated with conifer plantations mainly in western Britain; it may ultimately occur in Surrey, so careful checking of specimens is essential. The larvae are associated with wet decaying timber. All three *Sphegina* species known from Surrey do occur together, but this is seemingly rare as only three tetrads are common to all of them. The most likely reason is that occasions when this genus is abundant are relatively unusual, and so opportunities to collect sufficient specimens to secure all three species are limited. Even so, it would also appear that *S. clunipes* is far more ubiquitous and that the others have more demanding habitat requirements.

Sphegina clunipes (Fallén, 1816)

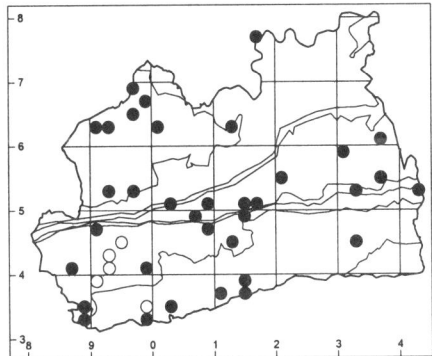

Number of records: 46

Surrey Status: Local

Flight times: April – September

Peak: June

This is the commonest of the three *Sphegina* species in Surrey. It is more widespread than the other two and occurs both on and off the Chalk. The best way to find this and the following two species is to look at the undersides and edges of umbel flowers such as wild angelica and hogweed in partially lit woodland locations.

Flower visits: garlic mustard, enchanter's nightshade

Sphegina elegans Schummel, 1843 = *kimakowiczi* (Strobl, 1897)

Number of records: 44
Surrey Status: Local
Flight times: May – August
Peak: June

This species can be detected by the same methods as those employed to find *S. clunipes*. This hoverfly is quite distinctive in the field as it has bright yellow humeri and clear abdominal markings. *S. elegans* is widely distributed on the sands and clays on either side of the Chalk, but is seemingly absent from the Chalk itself in Surrey.

Flower visits: cow parsley

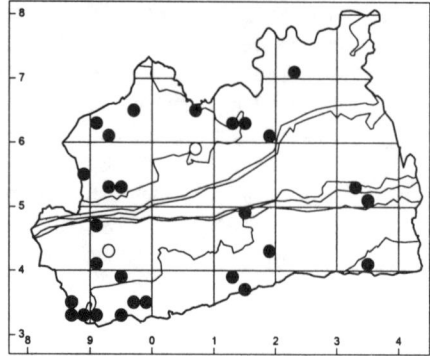

Sphegina verecunda Collin, 1937

Nationally Scarce
Number of records: 11
Surrey Status: Rare
Flight times: May – August
Peak: June

This scarce woodland hoverfly may be overlooked amongst the more numerous *S. clunipes* which it closely resembles. Although there are a limited number of records, this species seems to be concentrated off the Chalk, like *S. elegans*.

Flower visits: cow parsley, wild angelica

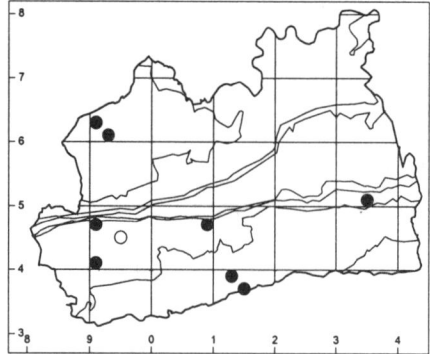

RECORDS: **Bagshot Plantations** SU9063 (29.5.1994, RKAM); **Thursley Common** SU9140 (5.6.1994, JSD); **Totford Lane Alder Wood** SU9147 (15.8.1992, RKAM); **West End Common** SU9360 (15.6.1988, GAC); **Charterhouse Alderholt** SU9544 (8.6.1968, PJC); **Gomshall** TQ0847 (20.7.1989, RKAM); **Chapel Copse** TQ1238 (4.7.1987, RKAM/GAC); near **Paynes Green** TQ1537 (7.6.1987, 14.6.1987, GAC); near **Godstone** TQ3550 (9.6.1996, RKAM).

Eristalini

This tribe comprises a range of wetland species; the larvae of some are truly aquatic and are the familiar "long-tailed maggots". Other larvae are associated with rotting leaf-sheaths of bulrush, whilst a few such as *Myathropa* and *Mallota* inhabit water-filled rot-holes. Many are very abundant at the flowers of hogweed and ragwort.

Genus *Anasimyia*

These are medium-sized yellow-and-black species with bars on the abdomen and grey stripes on the thorax, some of which are very similar. Of the five British species, three still occur in Surrey and one is thought to be extinct. The final species, *A. interpuncta,* is unlikely to occur in Surrey, but early examples of *A. contracta* and *A. transfuga* should be checked carefully as this species has been found away from its more traditional haunts in Cambridgeshire and the Thames marshes, and might be found if suitable habitat exists. The larvae are associated with decaying wetland vegetation, especially that of bulrush, *Typha* spp., and with marshes of reed sweet-grass, *Glyceria maxima*, and branched bur-reed, *Sparganium erectum*. The adults occasionally visit flowers such as yellow iris, *Iris pseudacorus*, but are more frequently seen flying through pondside vegetation.

Anasimyia contracta Claussen & Torp, 1980 PLATE 13

Number of records: 50
Surrey Status: Local
Flight times: May – August
Peak: June

This species is widely distributed and not uncommon at the margins of ponds where there is lush emergent vegetation including bulrushes. All specimens of this species should be carefully scrutinised for *A. transfuga*.

Flower visits: buttercup spp., water-cress

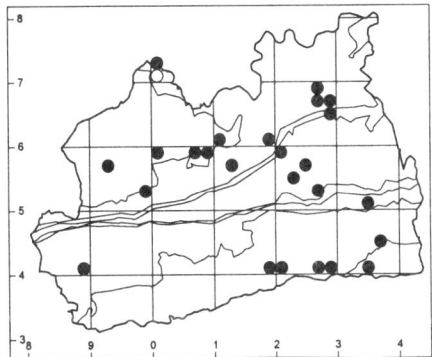

Anasimyia lineata (Fabricius, 1787)

Number of records: 69

Surrey Status: Local

Flight times: May – September

Peak: June

This is the commonest member of the genus in Surrey and can often be found in quite restricted habitats such as *Typha*-filled roadside ditches.

Flower visits: meadow buttercup, buttercup spp., water-plantain

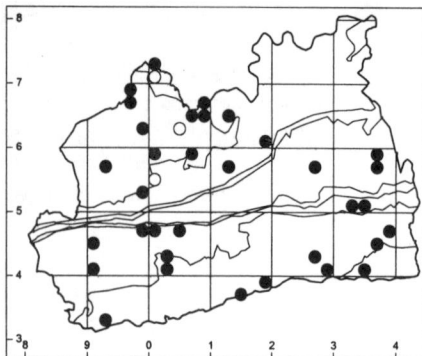

Anasimyia lunulata (Meigen, 1822)

Nationally Scarce

Number of records: 2

Surrey Status: Extinct

Flight times: June

This is mainly a western species which occurs widely in Wales. There are only two records from Surrey, both from heathland. It is to be feared that this species has been lost from what were localities on the extreme eastern edge of its range. Changing hydrology means that neither site is as wet as it was, and it seems unlikely that *A. lunulata* has survived. As part of a Biodiversity Action Plan, it would be worthwhile organising a search for this species at its former localities and elsewhere on wet heath.

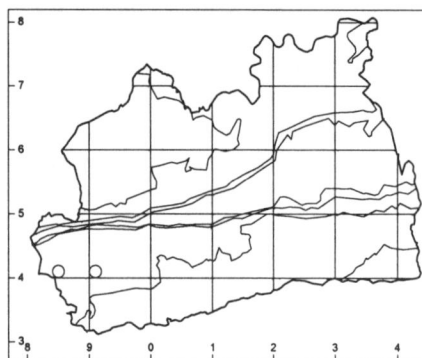

RECORDS: **Frensham Little Pond** 18.6.1938 (Diver Collection, ITE Furzebrook); **Thursley Common** SU9040 (2.6.1968, edge of cleared ditch, AES).

Anasimyia transfuga (Linnaeus, 1758)

Number of records: 5

Surrey Status: Rare

Flight times: May – June

Peak: May

There are very few records of this uncommon fly. Unlike its close relative *A. contracta*, *A. transfuga* appears to be more closely associated with marshy situations where reed sweet-grass, *Glyceria maxima,* is present. Open water does not seem to be a necessity and localities with seepage zones appear to

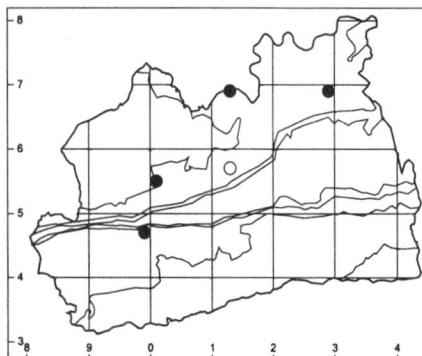

be favoured in west Surrey. I have ignored records of this species prior to 1983 when the separation from *A. contracta* became universally known.

RECORDS: **Shalford** SU9947 (7.5.1990, GAC); **Send Grove** TQ0154 (8.5.1993, RKAM); **Great Bookham Common** TQ1256 (10.6.1979, AES); Walton-on-Thames TQ1268 (21.5.1995, RKAM); **Mitcham Common** TQ2868 (1990, DCL).

Genus *Eristalinus*

These are medium to large flies with a distinctive metallic appearance to the abdomen, but no obvious markings. In build, they are squatter than other related genera. The larvae are associated with rotting vegetation.

Eristalinus sepulchralis (Linnaeus, 1758)

Number of records: 185

Surrey Status: Common

Flight times: April – September

Peak: August

This widely distributed hoverfly can be found at composite flowers throughout the summer. Most records are from the London Clay and Low Weald; my general impression is that this species is more commonly encountered in the London suburbs near sewage farms and rivers with a high level of sewage effluent. Elsewhere it is likely to be influenced by the level of livestock production. There are few records from the chalk where wetland is restricted to areas of overlying clays.

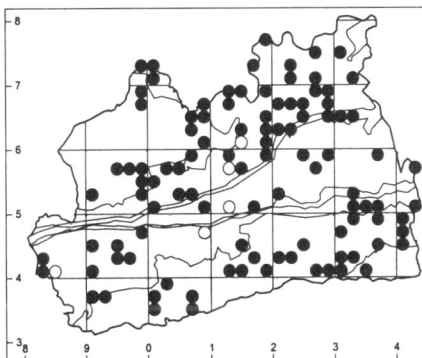

Flower visits: meadow buttercup, buttercup spp., water crowfoot (LP, 1950a), greater stitchwort, water-cress, wild cabbage, bastard cabbage, silverweed (LP, 1950a), tormentil (LP, 1950a), dogwood, wild parsnip (LP, 1950a), hemlock water-dropwort, hogweed (LP, 1950a), gipsywort (LP, 1950a), water mint, eyebright, prickly sow-thistle (LP, 1950a), *Hieracium* spp., common fleabane, Michaelmas-daisy, yarrow (LP, 1956b), scentless mayweed, common ragwort, hoary ragwort, ragwort spp., water-plantain

Eristalinus aeneus (Scopoli, 1763)

Number of records: 1

Surrey Status: Extinct

Verrall (1901) reports taking a specimen of this species from Denmark Hill on 14.8.1868. This is surprising because *E. aeneus* is normally a coastal species whose larvae occur in pools containing decaying seaweed (Rotheray, 1993). Given the origin of the record, it must be accepted as valid. At the time of this record, the tidal Thames was probably less heavily constrained by flood defences and this may have meant local breeding and migration, but it is equally possible that this was a vagrant from the north Kent coast.

Genus *Eristalis*

These are characteristic members of the so-called hogweed fauna, and most species might be expected to occur together on good sites. They are bee-mimics, some of which, such as *E. tenax,* show a close resemblance to the hive bee, *Apis mellifera.* The larvae are associated with wet decaying vegetation, especially around pond margins.

[*Eristalis abusivus* Collin, 1931

Number of records: 3

Surrey Status: Unconfirmed

This is normally a coastal species, but is known to occur inland. It is possible that it is overlooked inland because of its known coastal association, and so the records submitted to this project should not be dismissd outright. Unfortunately, the records noted here came to my attention too late to investigate them further; they are not listed for this reason.]

Eristalis arbustorum (Linnaeus, 1758)

Number of records: 689

Surrey Status: Ubiquitous

Flight times: April – October

Peak: August

A widely distributed and common hoverfly which is most frequent at both umbellifer and composite flowers. There is an interesting note by Groves (1956) recording the capture of a female at Carshalton which laid 82 eggs in captivity.

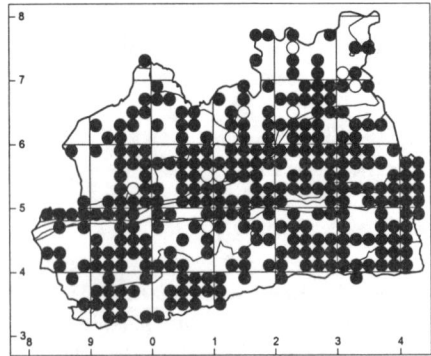

Flower visits: buttercup spp., creeping buttercup, greater stitchwort (LP, 1950a), redshank, Japanese knotweed, sallow (LP, 1950a), garlic mustard, water-cress, perennial wall-rocket, meadowsweet, bramble (LP, 1950a), silverweed (LP, 1950a), blackthorn (LP, 1950a), rowan, hawthorn, cow parsley, ground-elder, hemlock water-dropwort, wild angelica, wild parsnip, hogweed, upright hedge-parsley, wild carrot, ground-ivy, wild marjoram, water mint, eyebright, common valerian, devil's-bit scabious (LP, 1950a), creeping thistle, common knapweed (LP, 1956b), cat's-ear (LP, 1956b), rough hawkbit, perennial sow-thistle, dandelion, *Hieracium* spp., smooth hawk's-beard, common fleabane, Canadian goldenrod, Michaelmas daisy, sneezewort (LP, 1950a), yarrow, oxeye daisy, scentless mayweed, common ragwort, hoary ragwort, ragwort spp.

Eristalis horticola (De Geer, 1776)

Number of records: 143
Surrey Status: Common
Flight times: April – September
Peaks: May, <u>August</u>

During the early stages of the mapping scheme it seemed that this species' distribution was clumped to the east and west of the county. In recent years the gap through central Surrey has closed, but one still gets the impression that *E. horticola* has some underlying and undefined habitat requirements.

Flower visits: hawthorn, wild angelica, water mint, privet, devil's-bit scabious (LP, 1941), oxeye daisy, ragwort spp.

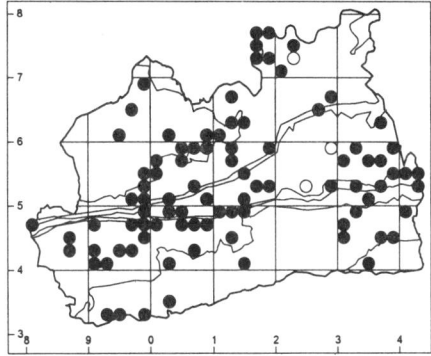

Eristalis interruptus (Poda, 1761) = *Eristalis nemorum*: auctt., misident.PLATE 15

Number of records: 352
Surrey Status: Ubiquitous
Flight times: April – September
Peak: August

This widely distributed and common hoverfly can be seen at composite and umbel flowers in most localities. Given the level of recording in the London area, it is surprising that this species is apparently scarce here, and it may be one of those hoverflies with a distinct negative correlation with the urban environment.

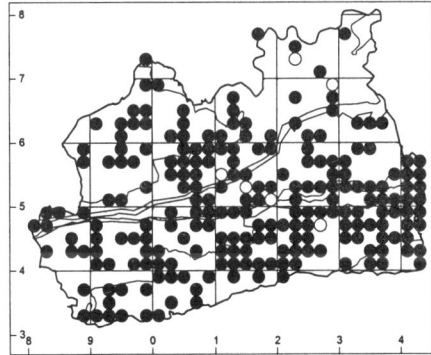

Flower visits: creeping buttercup (LP, 1950a), buttercup spp., garlic mustard, water-cress, meadowsweet, silverweed (LP, 1950a), hawthorn, ground-elder, hemlock water-dropwort, wild angelica, wild parsnip, hogweed, upright hedge-parsley, wild carrot, wild marjoram, corn mint, water mint (LP, 1950a), butterfly-bush, privet, devil's-bit scabious, creeping thistle, common knapweed, common fleabane, Canadian goldenrod, yarrow, oxeye daisy, scentless mayweed, common ragwort, ragwort spp., hemp-agrimony

Eristalis intricarius (Linnaeus, 1758)

Number of records: 185

Surrey Status: Common

Flight times: March – September

Peak: August

This is a widely distributed hoverfly which might be expected to occur at nearly any site, but especially in marshy locations. It is, however, seen sporadically but most commonly in the late summer.

Flower visits: bramble, blackthorn (LP, 1950a), hawthorn, wild marjoram, water mint, eyebright, devil's-bit scabious, small scabious, marsh thistle (LP, 1950a), creeping thistle, common knapweed, Michaelmas-daisy, scentless mayweed, hoary ragwort

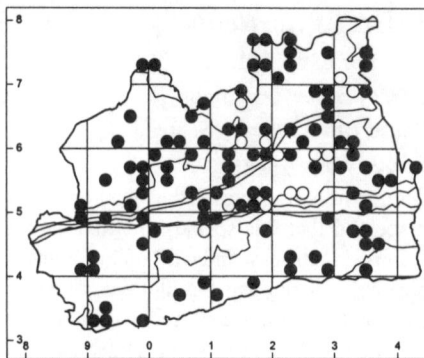

Eristalis pertinax (Scopoli, 1763)   PLATE 1

Number of records: 1,303

Surrey Status: Ubiquitous

Flight times: March – October

Peaks: May, August

This is one of the commonest hoverflies in the county which is most abundant in woodland at the flowers of umbellifers such as hogweed. There is a record of a single female caught in an insectocutor in a stable block in Richmond Park in 1994.

Flower visits: marsh marigold, creeping buttercup, buttercup spp., lesser celandine (LP, 1950a), Japanese knotweed, goosefoot spp., redshank, water-cress, heather, meadowsweet, bramble, dog rose (LP, 1950a), tormentil (LP, 1950a), blackthorn, wild cherry, cherry laurel, rowan, firethorn, hawthorn, wood spurge (LP, 1957), Norway maple, ivy, cow parsley, ground-elder, hemlock water-dropwort, wild angelica, wild parsnip, hogweed, upright hedge-parsley, wild carrot, wild marjoram, water mint, garden privet, *Hebe* sp., devil's-bit scabious (LP, 1950a), small scabious, marsh thistle (LP, 1950a), creeping thistle, perennial sow-thistle, dandelion, common fleabane, Canadian goldenrod, yarrow, oxeye daisy, scentless mayweed, common ragwort, hoary ragwort, ragwort spp., hemp-agrimony, water-plantain, ramsons

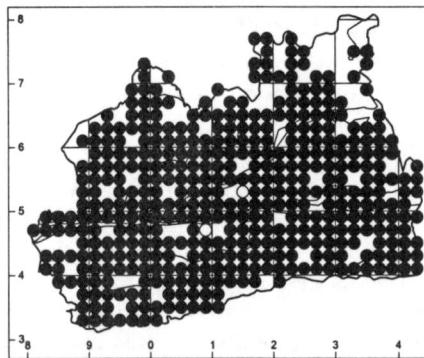

Eristalis tenax (Linnaeus, 1758)

Number of records: 767
Surrey Status: Ubiquitous
Flight times: January – December
Peak: August

This is obviously a widely distributed hoverfly, but it is likely to have been under-recorded because it is most frequent at the end of the summer when conditions for recording are uninspiring. Adults have been seen hibernating in holes in the walls of the fort at Box Hill, with many individuals packed close together (Morris & Collins, 1991). There is also an amusing account of a specimen of this species attracted to painted red flowers on wallpaper in a house in South Norwood which continued its mistake by visiting other similar painted bunches of flowers (Müller, 1872). In 1994 I examined a sample from an insectocutor in a stable block in Richmond Park which yielded 14 males and 182 females out of 214 hoverflies; as this species is known to be associated with organically rich wet sites (Stubbs & Falk, 1983), it seems possible that it is attracted to horse manure where it may breed if the stables are not cleaned out regularly.

Flower visits: creeping buttercup (LP, 1950a), buttercup spp., Japanese knotweed, perforate St John's-wort, imperforate St John's-wort, sallow (LP, 1950a), perennial wall-rocket, rape, *Erica* spp., dropwort, meadowsweet, silverweed (LP, 1950a), bird cherry, firethorn, hawthorn, bramble (LP, 1950a), dogwood, ivy, burnet-saxifrage, ground-elder, hemlock water-dropwort, wild angelica, wild parsnip, hogweed, wild carrot, field bindweed, forget-me-not, wild marjoram, water mint, butterfly-bush, *Hebe* sp., eyebright, devil's-bit scabious, small scabious, marsh thistle, creeping thistle, common knapweed, cat's-ear, autumn hawkbit, rough hawkbit, bristly oxtongue, perennial sow-thistle, prickly sow-thistle (LP, 1950a), dandelion, *Hieracium* spp., common fleabane, Canadian goldenrod, Michaelmas-daisy, sneezewort, yarrow, oxeye daisy, scentless mayweed, common ragwort, hoary ragwort, ragwort spp.

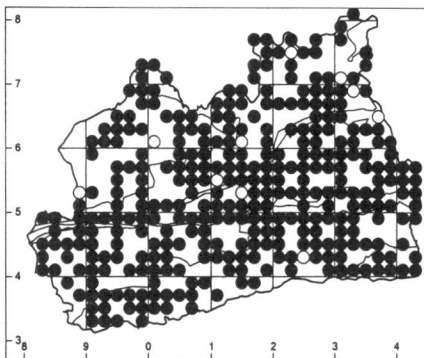

Genus *Helophilus*

These are distinctive yellow-and-black species with a striped black-and-grey thorax. Again, they are associated with wet decaying vegetable matter.

Helophilus hybridus Loew, 1846   PLATE 13

Number of records: 73
Surrey Status: Local
Flight times: May – September
Peak: August

Although widely distributed with scattered records across the county, *H. hybridus* is not common and records suggest that it may have declined somewhat, especially on the Chalk. It is mainly found in wetland locations, especially at the vegetated margins of ponds.

Flower visits: dogwood, wild marjoram, water mint, creeping thistle, sneezewort (LP, 1950a)

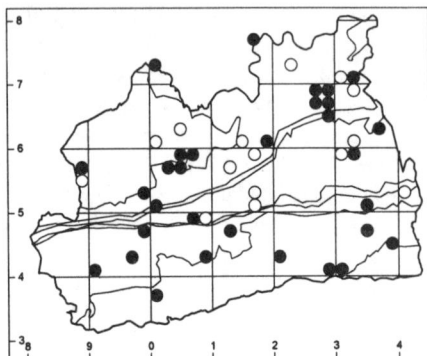

Helophilus pendulus (Linnaeus, 1758)

Number of records: 780
Surrey Status: Ubiquitous
Flight times: April – October
Peaks: May, <u>August</u>

This common hoverfly occurs thoughout the season and in almost any location.

Flower visits: creeping buttercup (LP, 1950a), buttercup spp., lesser spearwort (LP, 1950a), heather, *Erica* spp., meadowsweet, bramble, silverweed (LP, 1950a), creeping cinquefoil, dog rose, rowan, hawthorn, square-stalked willowherb, dogwood, ivy, cow parsley, ground-elder, hemlock water-dropwort, wild parsnip, hogweed, field bindweed, ground-ivy (LP, 1950a), wild marjoram, water mint, devil's-bit scabious, creeping thistle, common knapweed, bristly oxtongue, hawkweed oxtongue, perennial sow-thistle, prickly sow-thistle (LP, 1950a), dandelion, common fleabane, Canadian goldenrod, Michaelmas-daisy, yarrow, oxeye daisy, scentless mayweed, common ragwort, hoary ragwort, Oxford ragwort, ragwort spp.

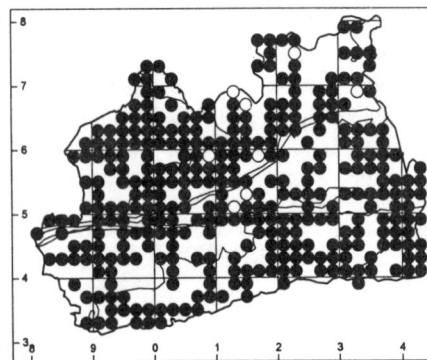

Helophilus trivittatus (Fabricius, 1805)

Number of records: 55
Surrey Status: Local
Flight times: May – September
Peak: August

The habitat preferences of this hoverfly are difficult to identify. It is rarely seen in numbers and often occurs in dry habitats, especially ruderal grasslands. Most records are from the high summer of August and September which might help to confirm the suggestion that the species is migratory (J. S. Denton, *pers. comm.*).

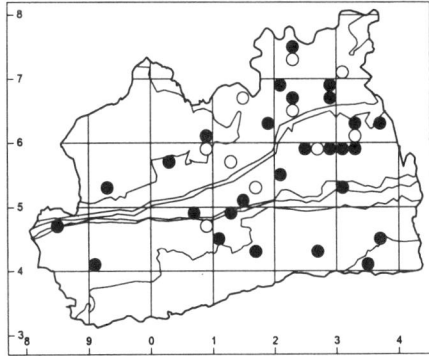

Flower visits: wild carrot, wild marjoram, devil's-bit scabious, creeping thistle, common knapweed (LP, 1956b), yarrow (LP, 1956b), oxeye daisy, hoary ragwort, ragwort spp.

Genus *Mallota*

Mallota cimbiciformis (Fallén, 1817)

Nationally Scarce
Number of records: 6
Surrey Status: Rare
Flight times: June – August
Peak: July

There are no recent records of this spectacular hoverfly, but the absence of such records cannot be used to suggest that it has been lost from the Surrey fauna. This is because it is probably easier to locate the larvae, and searching for them needs to be undertaken in appropriate localities which have large old

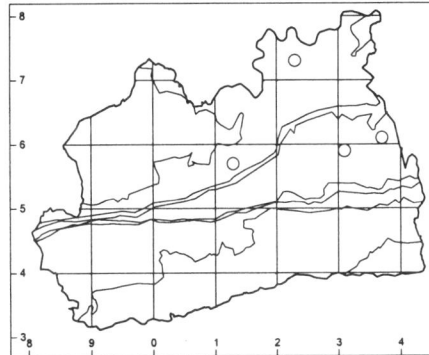

trees with substantial rot-holes. The most recent records are of larvae from rot-holes on Wimbledon Common (Jones, 1954).

Conservation: This is one of the assemblage of hoverflies which inhabit wet rot-holes in ancient trees, and site managers should be aware of the importance of such habitat and its retention. Creating new rot-holes by causing damage to younger trees may prove to be a useful activity. There is a need to survey for this species to determine its true status.

Flower visits: wild parsnip (LP, 1950a)

RECORDS: **Great Bookham Common** (11.8.1946, LP); **Wimbledon Common** (larvae 25.3.1951, 11.3.1952, AWJ); **Coulsdon** (27.6.1948, 3.7.1948, COH); **Selsdon Wood** TQ3661 (6.7.1942, RLC).

Genus *Myathropa*

Myathropa florea (Linnaeus, 1758)   PLATE 6

Number of records: 795

Surrey Status: Ubiquitous

Flight times: April – October

Peak: August

This is a very widely distributed hoverfly whose larvae can be found in just about any water-bearing depression in a tree. Males have been observed defending sunlit spots on leaves (Morris, 1991). This species can be quite variable in colour, with well-marked specimens being quite spectacular and poorly-marked specimens requiring a second check to confirm their identity. There is a record of a single female caught in an insectocutor in a stable block in Richmond Park in 1994 (see *E. tenax,* page 151).

Flower visits: lesser spearwort, knotgrass, bramble, firethorn, hawthorn, dogwood, holly, ivy, wood spurge (LP, 1957), cow parsley, burnet-saxifrage, ground-elder, hemlock water-dropwort, hemlock, wild angelica, wild parsnip, hogweed, upright hedge-parsley, wild carrot, field bindweed, wild marjoram, gipsywort, water mint, eyebright, devil's-bit scabious, creeping thistle, rough hawkbit, perennial sow-thistle, *Hieracium* spp., common fleabane, Canadian goldenrod, Michaelmas-daisy, yarrow (LP, 1956b), scentless mayweed, hoary ragwort, ragwort spp., water-plantain

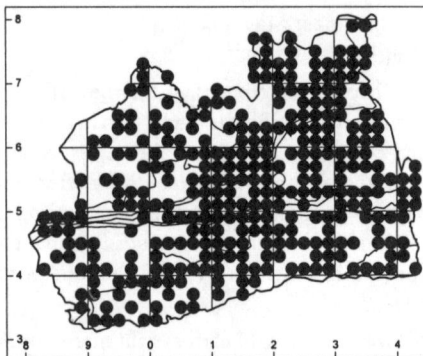

Genus *Parhelophilus* PLATE 13

Although generally smaller than *Helophilus*, there are occasions when small individuals of *H. pendulus* can be mistaken for this genus until the antennal colour is checked. The adults usually frequent pond-side vegetation and the larvae are associated with decaying emergent vegetation, especially that of bulrushes, *Typha* spp. I have also seen specimens of this genus with two black rings on the hind tibiae, suggesting that they should key to *Anasimyia*; inexperienced recorders should therefore retain vouchers of these two groups until they are thoroughly familiar with them. *Parhelophilus frutetorum* and *P. versicolor* are found cumulatively in some 60 tetrads, but only 10 (17%) are common to both. This suggests that there are some considerable differences in habitat requirements, but at the moment it is not possible to identify the factors which separate the two. As I have indicated below, it seems likely that pH is a factor, but more work is required to confirm or refute this suggestion.

Parhelophilus frutetorum (Fabricius, 1775)

Number of records: 55
Surrey Status: Local
Flight times: May – July
Peak: June

This is a widely distributed wetland hoverfly which is mostly found amongst rank vegetation at the edge of ponds and ditches. It is sometimes found some considerable distance from wetland sites which may suggest that it is quite mobile. A high proportion of records seem to be from the Lower Greensand and the Bagshot Sands, perhaps suggesting an association with more acid situations.

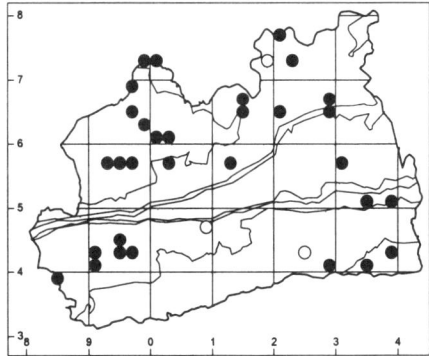

Parhelophilus versicolor (Fabricius, 1794)

Number of records: 65

Surrey Status: Local

Flight times: May – August

Peak: June

Like *P. frutetorum*, this species is most frequently found amongst rank vegetation at pond margins and amongst rank ditch vegetation. There would appear to be a close association with bulrushes, *Typha* spp. It is slightly commoner than *P. frutetorum*, but seemingly less common on the Bagshot Sands and the Lower Greensand, and almost absent from much of south-west Surrey.

Flower visits: water-cress, wild cabbage, cow parsley, hemlock water-dropwort, hogweed, field bindweed, yellow iris

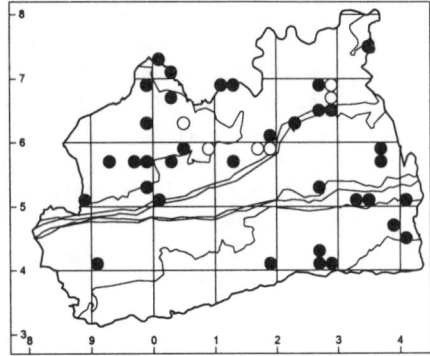

Merodontini

In Britain this tribe is represented by just six species, but in the hotter Mediterranean climate, genera such as *Merodon* and *Eumerus* are extremely diverse. Two species, *Merodon equestris* and *Eumerus tuberculatus*, are introductions from the continent in imported bulbs, and are considered horticultural pests. The biology of the other genus, *Psilota*, is extremely poorly known, and it may not actually be related to either *Merodon* or *Eumerus* (F. Gilbert, *pers. comm.*). All six species are known from Surrey, although only two are common.

Genus *Eumerus*

This genus includes three species of small black wasp mimics which are very similar and could cause confusion; specimens should therefore be retained as vouchers for critical examination. The larvae tunnel the rhizomes and bulbs of a variety of plants.

Eumerus ornatus Meigen, 1822

Nationally Scarce
Number of records: 17
Surrey Status: Scarce
Flight times: May – August
Peaks: <u>June</u>, August

This woodland hoverfly appears to be closely associated with older and possibly ancient woodland. There are inconsistencies however, with records from the Oxshott area. This hoverfly is most frequently seen flying or on foliage in dappled sunlight and is quite distinctive on account of its size, shape and abdominal markings. *E. ornatus* is listed on the long list of the UK Biodiversity Action Plan (DoE, 1995) and as such may be a candidate for special action in Surrey. At the moment, so little is known about its biology that no obvious action can be taken except further survey and possibly autecological studies. Without vouchers to refer against, this species can be difficult to identify, and I have frequently seen specimens of *E. tuberculatus* labelled as *E. ornatus* in collections, but once seen there can be little doubt about its identity.

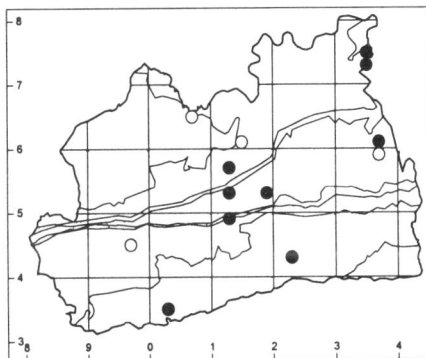

RECORDS: **Farncombe** (3.8.1923, AT); **Weybridge** (No date, GHV); **Fir Tree Copse** TQ0235 (23.6.1996, GAC); **White Downs** TQ1249 (26.6.1994, AJH); **White Hill** TQ1252 (11.6.1994, RKAM); **Effingham** (26.8.1934, LP, 1935); **Great Bookham Common** (13.8.1972, PJC), TQ1256 (6.6.1988, GAC); **Oxshott Heath** (1.7.1941, LP); **Headley Heath** TQ1953 (16.6.1995, ME); **Edolphs Copse** TQ2343 (14.6.1987, RKAM); **Sydenham Hill Wood** TQ3472 (1988, AG); **One Tree Hill** TQ3574 (25.8.1995, RAJ); **Selsdon Wood** TQ3661 (1.8.1930, 5.7.1931, RLC), (28.5.1990, RKAM); **Chelsham** (1942, RLC).

Eumerus strigatus (Fallén, 1817)

Number of records: 11
Surrey Status: Rare
Flight times: May – September
Peak: August

This is a remarkably scarce hoverfly which may be overlooked on account of its similarity to *E. tuberculatus,* which is much commoner. On a national scale, this insect appears to be most common in East Anglia, often in wetland locations. Known larval foodplants include *Iris* and this may explain its possible preference for wetter sites in parts of eastern England (Ball & Morris, *in press*), but some sites such as Albury Heath are dry, so a wider range of host plants must be used.

RECORDS: **Thursley Village** SU9040 (1.5.1992 - 31.7.1992, JSD); **Tuesley** SU9642 (15.7.1989, GAC); **Hillside House, Godalming** (16.6.1929, CD); **Albury Heath** TQ0646 (16.8.1987, RKAM/GAC), (7.1997, JSD); **Wisley Common** TQ0658 (3.9.1978, AES); **The Sheepleas** TQ0851 (16.6.1991, AJH); **Reffolds Copse** TQ1843 (7.6.1987, GAC); **Headley Heath** TQ2053 (18.8.1991, ME); **Blindley Heath** TQ3644 (11.8.1991, JRD).

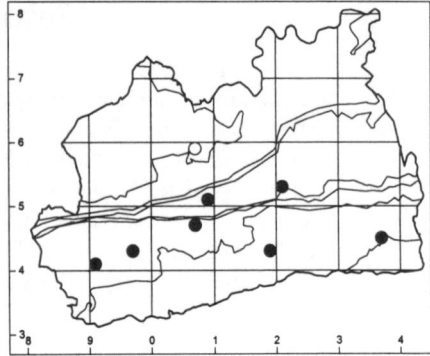

Eumerus tuberculatus Rondani, 1857 **The Lesser Bulb Fly**

Number of records: 129
Surrey Status: Local
Flight times: April – September
Peaks: May, <u>August</u>

This is a continental species which was almost certainly introduced to Britain with imported bulbs. It was not known at the time of Verrall, and was added to the British list by Collin (1918). Unlike some other notable additions to the British fauna, there does not seem to have been a rush of records arising from this species' announcement as new to Britain, and its spread in Surrey is not obviously documented. It is now a common, but perhaps under-recorded species which should occur throughout the county. The larvae are known to mine *Iris* and *Narcissus* bulbs (Stubbs & Falk, 1983) and it might be expected that it would be more commonly associated with the urban environment from which it originated; this seems to be the case in the London area where urban bulb-planting may supplement wild host-plant sources. There would also seem to be a higher density of records from well-drained areas such as the sands of west Surrey, while it is remarkably scarce in the Low Weald. The reasons for this apparent difference are unclear, but may be related to lower densities of

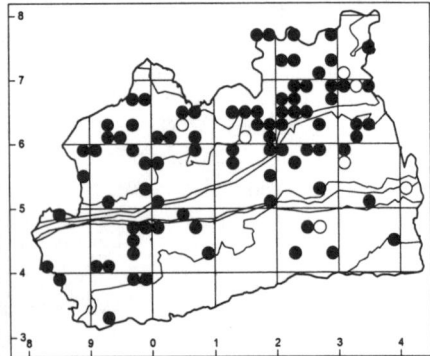

housing and related gardens with suitable bulbs, but it is also a possibility that *E. tuberculatus* cannot cope with conditions such as impeded drainage.

Flower visits: creeping buttercup, buttercup spp., lesser stitchwort, knotgrass, Japanese knotweed, tormentil, hawthorn, wood spurge (LP, 1957), wild carrot, field bindweed, smooth hawk's-beard

Merodon equestris (Fabricius, 1794) **The Greater Bulb Fly**   PLATE 14

Number of records: 271

Surrey Status: Common

Flight times: May – August

Peak: June

We can be reasonably sure of one origin of this continental species in the UK, for Verrall (1901) reports that specimens in the garden of his brother at Denmark Hill were the first British records and probably originated in Dutch bulbs which "were annually purchased by him for that garden, and I have scarcely any doubt but that the species was imported about that time". Since then, *M. equestris* has spread throughout much of the UK into northern Scotland. This is a widespread species in Surrey, although it is more frequent in the suburban landscape. All four colour forms depicted in Stubbs and Falk (1983) have been recorded, but no attempt has been made to map each morph separately.

Flower visits: buttercup spp., greater stitchwort, tormentil (LP, 1950a), hogweed, field bindweed, smooth hawk's-beard, beaked-hawk's-beard, oxeye daisy

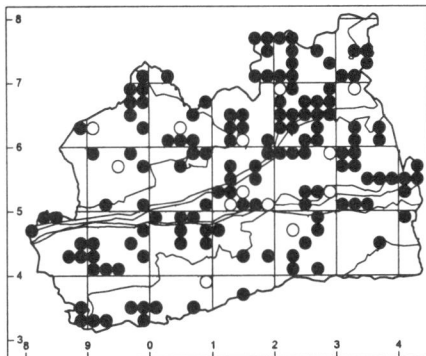

Genus *Psilota*

Psilota anthracina Meigen, 1822

Red Data Book 2

Number of records: 6

Surrey Status: Rare

Flight times: May – June

Peak: May

The records of this scarce hoverfly are perplexing as it appears to favour heathland and parkland equally. There is a single record of a specimen taken at an oak sap run (JSD), and the larvae are known to breed in sap runs (Rotheray, 1993). It is most frequently recorded at the flowers of hawthorn where it may be overlooked as a small calypterate. More frequently, however, small calypterates are captured with great excitement, only to be discarded with much disappointment.

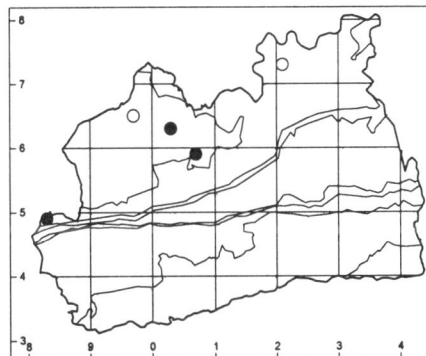

Flower visits: hawthorn

RECORDS: **Farnham Park** SU8348 (23.5.1997, JSD); **Chobham Longcross** SU9765 (3.6.1979, AES); **Woodham** TQ0362 (7.5.1988, GAC); **Wisley RHS Gardens** TQ0658 (16.5.1988, AJH); **Richmond Park** (3.6.1971, 26.5.1974, COH).

Pelecocerini

Two genera, *Chamaesyrphus* and *Pelecocera* are known in Britain, but the former is confined to Scotland. Very little is known about their larval biology.

Genus *Pelecocera*

Pelecocera tricincta Meigen, 1822

Red Data Book 3
Number of records: 4
Surrey Status: Rare
Flight times: June – Sept
Peak: August

The records of *P. tricincta* are the most northerly in the UK for this heathland species which is better known from the heaths and bogs of the New Forest and Dorset. *P. tricincta* may be more widespread on the Surrey heaths than the records suggest and should be searched for on small yellow composite flowers along heathland footpaths. It is perhaps noteworthy that Verrall's comment (1901) that this species "is almost certain to occur in the neighbourhood of Woking where so many of the Dorset and New Forest species re-occur" foresaw the discovery of this species in Surrey some 83 years later. I find it equally surprising that *P. tricincta* is apparently absent from the heaths of south-west Surrey; now meet the challenge!

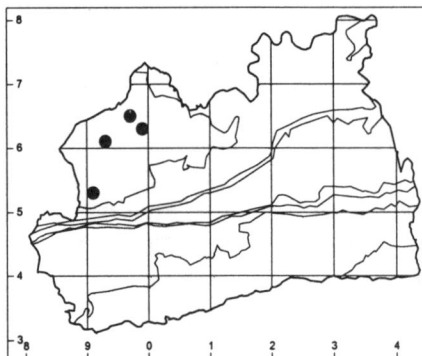

Conservation: The current range of *P. tricincta* in Surrey is confined to the larger tracts of heathland and this may be the key to its continued survival. All efforts to prevent and revert scrub and bracken invasion should be continued, and flowery heathland verges retained to provide nectar sources.

Flower visits: cat's-ear

RECORDS: **Ash Ranges - Fox Hills** SU9152 (26.8.1984, SRM), **Pirbright Ranges** SU9261 (6.8.1993, GM); **Chobham Common** SU9963 (27.6.1987, RKAM), SU9764 (17.9.1988, GAC).

Pipizini

This tribe comprises genera which are generally regarded as difficult to identify to species, and therefore many hoverfly enthusiasts ignore or avoid them. With patience, however, most species can be keyed out with reasonable confidence. The genitalia of the males should be extended to allow examination either at the time, or by future workers. Recording scheme organisers will generally be pleased to receive specimens of these genera. Formerly, female *Neocnemodon* were not thought to be identifiable, but can now be keyed through with some confidence using the key in Stubbs (1996).

Genus *Heringia*

This genus of small black hoverflies now also includes those species formerly in *Neocnemodon,* which has been reduced to a sub-genus. The adults are infrequently seen, but when found will often be noted sunning on leaves. The larvae are predators on a variety of arboreal aphids.

Heringia heringi (Zetterstedt, 1843)

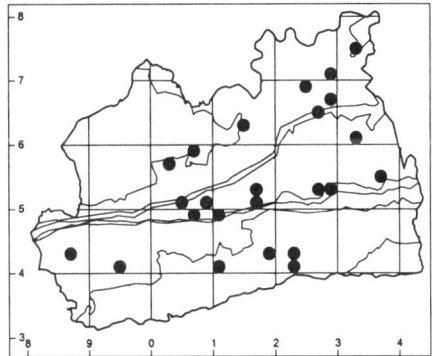

Number of records: 28
Surrey Status: Local
Flight times: May – August
Peak: May

This is a widespread hoverfly which can often be found sunning itself on sycamore and hawthorn leaves. Although it also visits flowers and has been recorded from apple in considerable numbers at Happy Valley, Box Hill, it usually occurs as single individuals. It does appear to be quite a variable species, especially in the degree of wing shading.

Flower visits: apple

Heringia brevidens (Egger, 1865)

Nationally Scarce

Number of records: 1

Surrey Status: Rare

The single record of this species from Willow Lane, Mitcham, on 24.4.1949 at marsh marigold (LP) was also the first for the UK (Stubbs, 1980). I have not been able to identify the site accurately, but believe that it was probably on the banks of the River Wandle where Willow Lane industrial estate now stands. *H. brevidens* is possibly associated with poplar aphids, having been taken by sweeping poplar leaves elsewhere (S. Falk, *pers. comm.*).

Flower visits: marsh marigold (Stubbs, 1980)

Heringia latitarsis (Egger, 1865)

Nationally Scarce

Number of records: 13

Surrey Status: Rare

Flight times: May – August

Peaks: <u>May</u>, August

Heringia latitarsis seems to be more or less confined to the sands of west Surrey where it has been found in woodland rides and in heathy situations. The reasons for this concentration is far from clear, but may be related to local micro-climate.

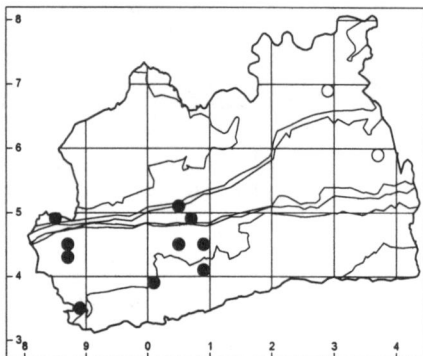

RECORDS: **Farnham** SU8548 (15.8.1992, RKAM); **Tilford Reeds** SU8643 (9.5.1987, GAC); **Waverley Abbey** SU8745 (9.5.1987, RKAM/GAC); **Hindhead Common** SU8934 (28.6.1987, RKAM); **Hascombe Hill** TQ0039 (29.8.1992, RKAM); above **Woodhill Sand Pit** TQ0444 (29.8.1992, RKAM); **Woodland near Clandon** TQ0450 (23.8.1992, RKAM); **Netley Heath** TQ0749 (4.6.1986, GAC); **Ewhurst Woods** TQ0840 (22.7.1995, RKAM); **Hurtwood** TQ0844 (8.5.1989, RKAM); **Mitcham Common** TQ2868 (5.1974, AES); **Chelsham** (6.7.1942, RLC).

Heringia pubescens (Delucchi & Pschorn-Walcher, 1955)

Nationally Scarce
Number of records: 5
Surrey Status: Rare
Flight times: April – June
Peak: April/May

Current records of this small black species are insufficient to make any significant observations on its distribution and ecology. The common factor at Roothill and Witley Common is the presence of conifers, but there is no such link at Great Bookham Common and Merrow Common.

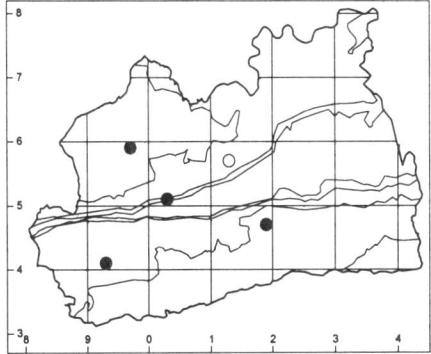

RECORDS: **Witley Common** SU9340 (16.5.1992, RKAM); **Barrs Lane, Knaphill** SU9659 (28.4.1991, AJH); **Merrow Common** TQ0251 (28.4.1990, PLTB); **Great Bookham Common** (17.6.1942, LP); **Roothill** TQ1947 (24.5.1986, RKAM).

Heringia vitripennis (Meigen, 1822)

Number of records: 22
Surrey Status: Scarce
Flight times: April – August
Peaks: May, July

In 1990/1991 I ran a series of water traps on Mitcham Common, two of which were in bramble patches where this species was taken in some numbers. This suggests that *N. vitripennis* could be associated with brambles and is possibly more widespread than the maps suggest.

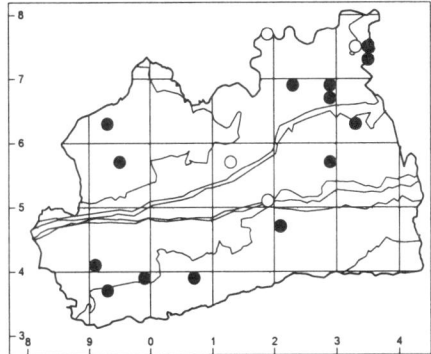

Genus *Pipiza*

These are small to medium-sized black species, many of which sport a pair of yellow spots on tergite 2. On a national scale, very few records are submitted for this genus even though some are remarkably common, being found sunbathing on leaves in the spring. They are, however, difficult to identify and this means that they are poorly recorded by most hoverfly enthusiasts. The larvae are aphid predators and are associated with both ground-layer and arboreal aphids.

Pipiza austriaca Meigen, 1822

Number of records: 62
Surrey Status: Local
Flight times: May – August
Peaks: June, August

This is one of the more straightforward species of the genus to identify and females can be provisionally identified in the field by the very grey appearance of the abdomen caused by the hair patterns on the tergites; its larvae are associated with woodland umbels (G. Rotheray, *pers. comm.*). Records are widely scattered and often come from wooded locations where adults can often be found at flowers as well as basking in the sun. I have a specimen in my collection from Banstead, TQ2460, on 16.6.1987 (RKAM) which appears to be this species, but for yellow spotting on tergite 2.

Flower visits: hawthorn, hogweed

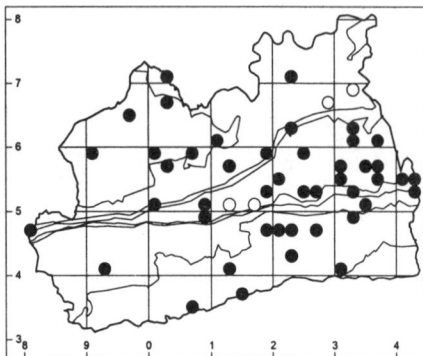

Pipiza bimaculata Meigen, 1822

Number of records: 17
Surrey Status: Rare
Flight times: April – August
Peak: May

This is one of the most difficult species to identify and also one which is very similar to *P. noctiluca* form F (Stubbs & Falk, 1983). It is the most likely species within this genus to have been overlooked. Current records give no clear clue of habitat preferences; most are from April and May, but there are two from July/August which suggests a partial second generation. All records of this species must be treated with distinct caution, including those published here.

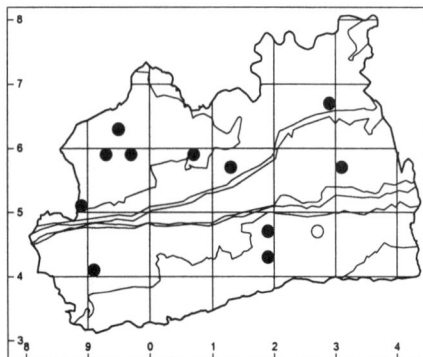

Pipiza fenestrata Meigen, 1822

Number of records: 16

Surrey Status: Rare

Flight times: May – September

Peak: May/June

This is a particularly difficult species to identify and trustworthy records are scarce. In the field, large *Pipiza* specimens suggest that this species has been found, but I have never found the critical characters easy to confirm, i.e. its taxonomic status is uncertain. All records, bar one, are from May and June.

Flower visits: creeping buttercup (LP, 1960)

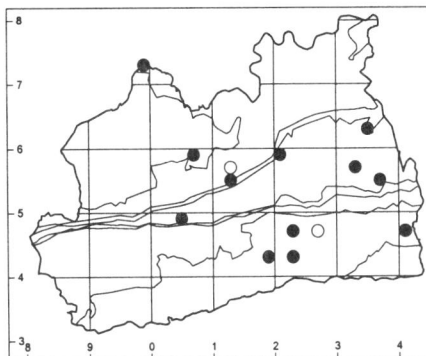

Pipiza lugubris (Fabricius, 1775)

Nationally Scarce

Number of records: 11

Surrey Status: Rare

Flight times: July – August

Peak: August

Many of the sites where I have encountered this hoverfly support large quantities of meadowsweet and it is possible that this is one clue to the distribution of *P. lugubris* (Morris, 1993). Female *P. lugubris* are very distinct with a chocolate-brown wing smudge quite unlike other species of the genus, but I have two males in my collection that fit all of the characters of *P. lugubris* except that they have almost clear wings. Beuk (1989) also records taking a pair of this species where the male had almost clear wings.

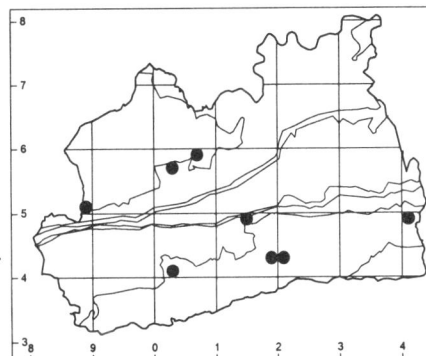

RECORDS: **Ash Wharf** SU8951 (15.8.1987, GAC); **Woking** (specimen in Hope Collection, no data, col. "HGC" - possibly H.G. Champion); **Cartbridge** TQ0256 (22.7.1989, mf., PLTB); **Run Common** TQ0341 (16.8.1987, RKAM); **Wisley RHS Gardens** TQ0659 (1982, 08.8.1997, AJH); **Wisley Common** (3.7.1965, AES); **Milton Heath** TQ1548 (5.7.1988, GAC); **Reffolds Copse** TQ1843 (29.8.1987, GAC); **Hammonds Copse** TQ2143 (25.8.1985, RKAM); **Staffhurst Wood** TQ4148 (30.8.1987, RKAM).

Pipiza luteitarsis Zetterstedt, 1843

Number of records: 25
Surrey Status: Scarce
Flight times: April – May
Peak: May

During my initial years studying hoverflies I failed to find this species despite taking a good many specimens of *Pipiza*, and concluded it was not common. In 1990, however, it was widespread and common, therefore suggesting that it occurs in widely fluctuating numbers. The larvae of this species feed on *Schizoneura* aphids on elm (Rotheray, 1987) and males can be found defending sunny locations, especially in groves of elm scrub (Morris, 1991).

Flower visits: rhododendron

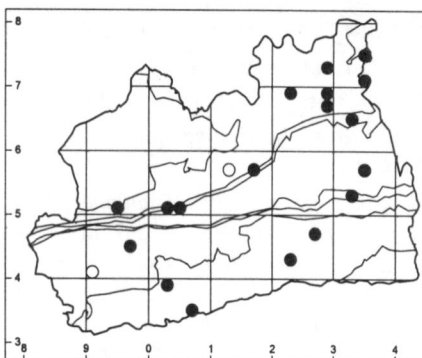

Pipiza noctiluca (Linnaeus, 1758)

Number of records: 150
Surrey Status: Common
Flight times: April – October
Peak: May

At the moment, this species is a dumping ground for a whole range of specimens with differing hair patterns and coloured spotting, and it is possible that more than one species is involved. Records are most common in the spring and early summer. In my collection there are more male form F than typical *noctiluca*, whilst the majority of females are typical *noctiluca* with only five form E. I also have two specimens that appear to be male form D, and three males with grey dusting on tergite 2 where yellow spots would occur in form F, which might be an additional form to those described in Stubbs & Falk (1983).

Flower visits: buttercup spp., greater stitchwort (LP, 1950a), wood spurge (LP, 1957), cow parsley, wild parsnip, hogweed, common ragwort

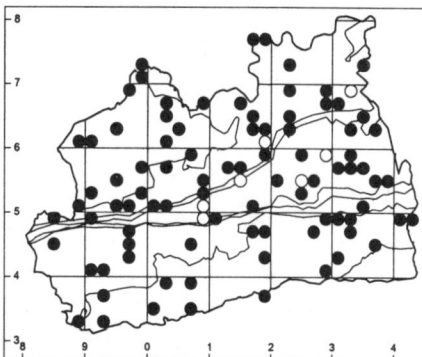

Pipiza spp. Four-spotted forms

I possess four specimens from Surrey in my collection with spots on tergite 3 as well as tergite 2. They are extremely variable in size, ranging from the size of *P. fenestrata* to almost that of *P. bimaculata*. The body is elongate and therefore this is likely to mean that they are not *P. quadrimaculata,* which is described by Speight (1988) as similar to *Trichopsomyia flavitarsis* but with females having "a short broad abdomen".

RECORDS: near **Pirbright** SU9555 (9.5.1992, f., RKAM); **Merrow Common** TQ0251 (28.4.1990, m., RKAM). **Cannon Hill Common** TQ2368 (2.5.1990, mm., RKAM).

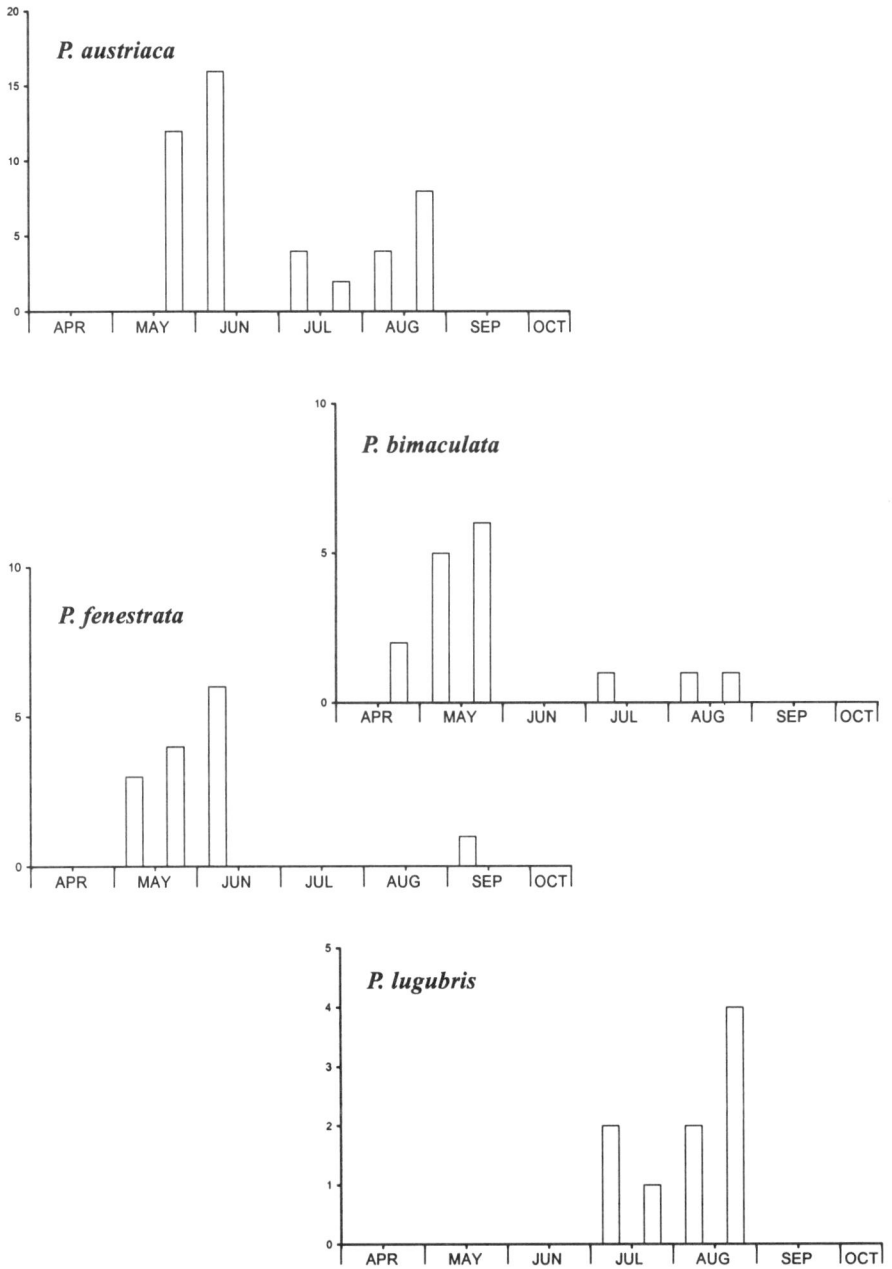

Figure 14a. **Phenology of *Pipiza* species**

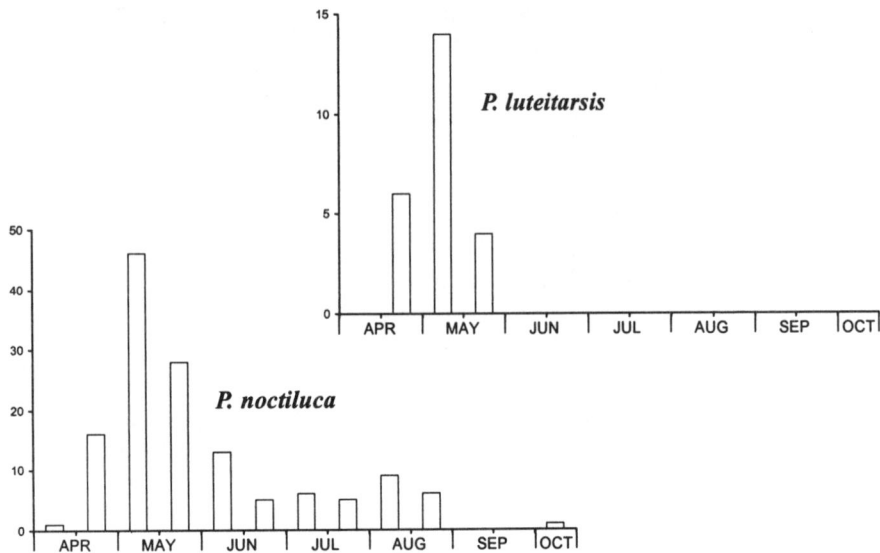

Figure 14b. **Phenology of *Pipiza* species**

Genus *Pipizella*

These are small dark species not dissimilar to *Pipiza*, but they do not posess yellow markings on tergite 2 and the wing venation differs substantially in the shape of the outer cross-vein. The larvae are associated with root-feeding aphids on umbellifers (Rotheray, 1993). Males can be quite readily identified once the genital capsule is extended, but females cannot be identified with such confidence until voucher material has been assembled. The use of the yellow aristal character to separate female *P. viduata* from *P. virens* in Stubbs & Falk (1983) is unreliable and a better character is the length of the hairs on the hind tibiae, once comparative material is available.

Pipizella maculipennis (Meigen, 1822)

Red Data Book 3

Number of records: 1

Surrey Status: Rare

There are very few reliable records of this hoverfly on a national scale, let alone from Surrey; in part, this may be because it is very difficult to identify with certainty as a female. One unconfirmed record is by Verrall (1901), who reports that "I possess a female which may belong to this species which I caught at Reigate on July 5th 1872".

RECORDS: **Wisley** TQ0558 (17.7.1965, AES).

Pipizella viduata (Linnaeus, 1758) = *varipes* (Meigen, 1822)

Number of records: 177
Surrey Status: Common
Flight times: April – September
Peaks: <u>June</u>, August

This widely distributed and common hoverfly possibly prefers drier localities and is much scarcer on the clays of the Low Weald. It can often be found at small umbels in grasslands and the larvae are known to be linked to root aphids in association with ants. Adults are frequently found sunning on bramble and other leaves, and are often numerous where they occur. I have in my collection a female from Epsom (TQ2064) on 10.6.1995 with yellow tarsi on the fore and mid legs; it is rather dumpier than normal *P. viduata*. At the moment this specimen cannot be identified with certainty, but may be *P. annulata*.

Flower visits: meadow buttercup, bulbous buttercup, buttercup spp., common rock-rose, tormentil, creeping cinquefoil, cow parsley, burnet-saxifrage, ground-elder, wild angelica, wild parsnip, hogweed, upright hedge-parsley, wild carrot, common marsh bedstraw, beaked hawk's-beard

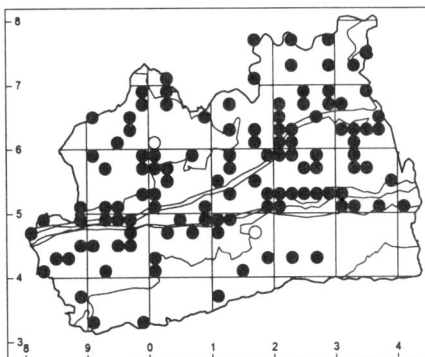

Pipizella virens (Fabricius, 1805)

Nationally Scarce
Number of records: 52
Surrey Status: Local
Flight times: May – August
Peak: June

The majority of records suggest that this is a grassland hoverfly which is widely distributed in Surrey but is considerably less common than *P. viduata*. Like *P. viduata* it appears to be more frequent in grassland and scrub edge on better drained soils on chalk and sand. It is really only possible to identify the genus on microscopic examination and with the aid of voucher material; males are relatively straightforward on genitalia characteristics once the genital capsule has been prepared, but identifications should not rely on the colour of the arista as many *P. viduata* also have a yellowish arista. *P. virens* is generally larger and slightly hairier than *P. viduata*, and the lengths of the hairs on the hind tibia are a particularly useful character once voucher material is available for comparison. All specimens should therefore be retained for critical examination.

Flower visits: buttercup spp., hogweed, wild carrot

RECORDS: **Frensham** SU8440 (20.7.1956, JAH); **Thundry Meadows** SU8944

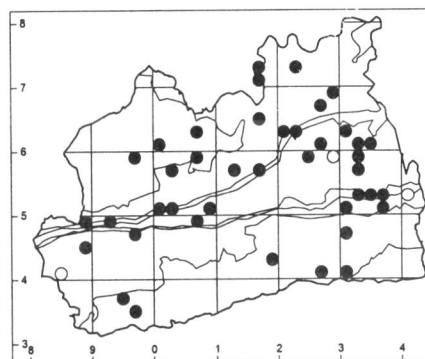

(10.8.1991, JRD); **Seale** SU8948 (11.7.1987, GAC); **Hog's Back** SU9248 (28.6.1987, RKAM); near **Witley Station** SU9537 (30.7.1994, RKAM); **Lion's Copse, Chiddingfold** SU9635 (18.5.1997, RDH); **Compton Common** SU9646 (7.8.1987, RKAM); **Barrs Lane, Knaphill** SU9659 (20.6.1991, AJH); near **Guildford** TQ0051 (20.5.1990, GAC); **Bonsey's Bridge, Horsell** TQ0061 (31.5.1997, GAC); **Merrow Common** TQ0251 (5.7.1987, GAC); **Cartbridge** TQ0256 (22.7.1989, PLTB); **Wisley RHS Gardens** TQ0658 (29.7.1988, AJH); **New Haw** TQ0663 (23.6.1995, RKAM); **Netley Downs** TQ0748 (5.8.1991, JRD); **The Sheepleas** TQ0851 (16.6.1988, RKAM); **Great Bookham Common** (pre-1950, CNC), TQ1256 (2.7.1986, 17.6.1987, GAC); near **Teddington** TQ1671 (23.6.1996, RKAM); **Eel Pie Island** TQ1672 (25.6.1989, RKAM); **Lower Ashtead** TQ1757 (18.6.1994, RKAM); **Claygate** TQ1764 (23.6.1995, RKAM); **Reffolds Copse** TQ1843 (7.6.1987, GAC); **Ewell** TQ2063 (10.6.1995, RKAM); **Nonsuch Park** TQ2262 (12.7.1985, VH); **Wimbledon Common** TQ2272 (27.5.1990, RKAM/JRD); **Ruffett Wood** TQ2558 (26.8.1994, RKAM); **Hookwood** TQ2641 (15.6.1994, RDH); **Riverside Park, Horley** TQ2741 (5.6.1994, RDH); **Banstead Downs** TQ2661 (18.6.1988, RKAM); **Mitcham - Watermeads** TQ2767 (26.6.1987, RKAM); **Mitcham Common** TQ2868 (18.7.1988, 23.7.1988, 28.8.1991, RKAM); **Fernhill** TQ3041 (4.7.1994, RDH); **Shepherd's Hurst** TQ3046 (8.8.1993, RKAM); **Nutfield Marsh** TQ3051 (5.7.1988, GAC); **Roundshaw** TQ3062 (24.7.1997, PG); **Coulsdon Common** TQ3257 (14.6.1995, RDH); **Coulsdon** (9.6.1964, LP); **Riddlesdown** TQ3260 (24.6.1986, GAC); **Riddlesdown Quarry** TQ3359 (10.6.1987, GAC); **Brewer Street** TQ3352 (19.7.1986, GAC); **Pilgrim Fort** TQ3453 (12.6.1987, GAC); **Sanderstead Plantation** TQ3461 (6.7.1988, GAC); near **Godstone** TQ3650 (1.6.1993, RKAM); TQ3652 (1993, CWP); **Limpsfield Common** (pre-1941, LP).

Genus *Trichopsomyia*

Trichopsomyia flavitarsis (Meigen, 1822)

Number of records: 10

Surrey Status: Rare

Flight times: May – July

Peak: June

This is mainly a northern hoverfly whose distribution in southern England is patchy. Its larvae are associated with the galls of the psyllid *Liva juncorum* on rushes, *Juncus* spp. (G. Rotheray, *pers. comm.*). In Surrey it shows a strong preference for acid localities and is clearly well-established at Thursley Common. There is a strong likelihood that this species is more widespread on wetter heathlands than current records suggest. The Richmond Park

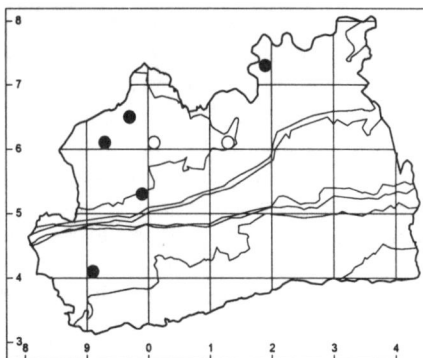

record, which was a single specimen taken in a rushy area, is most surprising and is the first for the London area (see Plant, 1986).

RECORDS: **Thursley Common** SU9041 (28.6.1972, unknown recorder), (22.7.1973, PJC), (12.7.1990, MAH/EAH), (14.7.1991, JRD); **Brentmoor Heath** SU9261 (24.6.1997, JSD); **Chobham Longcross** SU9765 (31.5.1992, AJH); **Whitmoor Common** SU9853 (15.6.1996, AJH); **Horsell Common** TQ0060 (22.7.1973, PJC); **Oxshott Heath** (11.6.1891, [Billups, 1891]); **Richmond Park** TQ1972 (6.6.1992, RKAM).

Genus *Triglyphus*

Triglyphus primus Loew, 1840

Nationally Scarce
Number of records: 8
Surrey Status: Rare
Flight times: May – August
Peak: August

This small and unassuming hoverfly may have been overlooked during the recent survey, especially as it is probably closely associated with ruderal situations such as roadside verges and waste ground where its larvae live in galls on mugwort, *Artemisia vulgaris*. My own experiences of this hoverfly are limited, but on two occasions I have found numerous individuals at wild parsnip on Mitcham Common.

Flower visits: wild parsnip

RECORDS: **St. Anne's Hill** TQ0267 (7.6.1986, GAC); **Chertsey Meads** TQ0665 (23.8.1987, GAC); **Wimbledon Common** (6.8.1949, COH); **Mitcham Common** (5.7.1947, LP), TQ2868 (1984, RDD/CMJ), (7.5.1989, PLTB), (27.8.1991, 28.8.1991, RKAM); **Thornton Heath** (30.6.1935, LP).

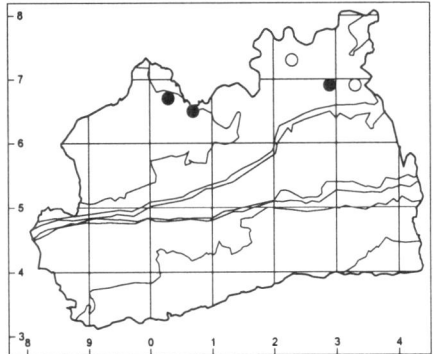

Sericomyiini

There are two genera in Britain, *Arctophila* and *Sericomyia*, both of which comprise mainly northern and western species (Ball & Morris, *in press*). Both species of *Sericomyia* occur in Surrey, favouring the wet acid conditions found on the heaths and nearby sandstones.

Genus *Sericomyia*

These two large hoverflies are quite distinctive and should not be easily misidentified. Their larvae are reported to be associated with peaty pools on moorlands (Rotheray, 1993) but, given their distribution in Surrey a wider range of possible wet habitats is likely to be utilised.

Sericomyia lappona (Linnaeus, 1758)   PLATE 10

Number of records: 33
Surrey Status: Scarce
Flight times: May – August
Peak: May

This is mainly a heathland species in Surrey with the exception of the population around Redlands Wood (TQ1545) which is indicated by records in 1989 and 1993. Further recording on the sands of the Hurtwood and Friday Street area may reveal a wider population nearby, as this species seems to be entirely associated with exposures of the Bagshot Sands and the Greensand.

Flower visits: buttercup spp., rowan

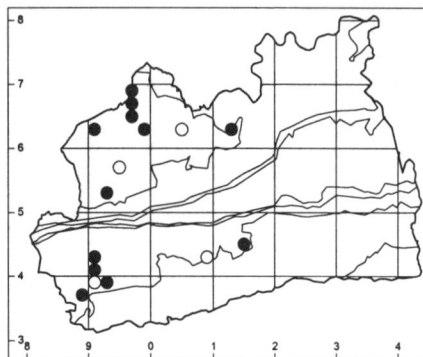

Sericomyia silentis (Harris, 1776)   PLATE 10

Number of records: 87
Surrey Status: Local
Flight times: May – October
Peak: September

There are scattered records of *S. silentis* from much of the county outside London. It is genuinely a scarce species which occurs most frequently on the heaths of west Surrey and in the woodlands on the edge of the North Downs and on the Greensand around Hurtwood.

Flower visits: bell heather, ivy, wild angelica (Uffen, 1969), upright hedge-parsley, devil's-bit scabious, cat's-ear, common ragwort, ragwort spp.

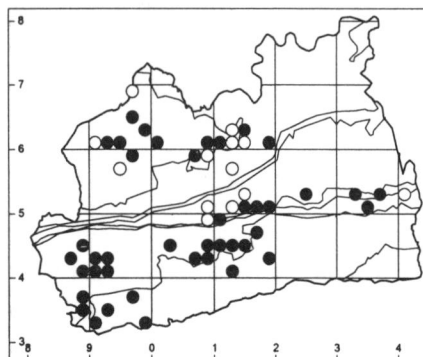

Volucellini

There is one genus in Britain comprising five species, of which all occur in Surrey. They are amongst the most spectacular of our large hoverflies, being mimics of bees, wasps and hornets. Four are known to be scavengers associated with the nests of bumblebees and social wasps, and one is associated with sap runs.

Volucella bombylans (Linnaeus, 1758)   PLATE 16

Number of records: 246
Surrey Status: Common
Flight times: May – August
Peak: June

This common hoverfly of late spring and early summer occurs typically in two forms, red-tailed (f. *bombylans*) and white-tailed (var. *plumata*), both of which are good bumblebee mimics. As no special recording effort has been made to separate the two forms, only the one map is presented here. On balance, however, personal observations suggest that var. *plumata* (37 records) is much commoner than the typical form (9 records). The larvae are reported to be associated with the nests of *Vespula germanica* (Stubbs & Falk, 1983), but are usually associated with those of bumblebees (G. Rotheray, *pers. comm.*).

Flower visits: bramble, wood spurge (LP, 1957), hogweed, water mint

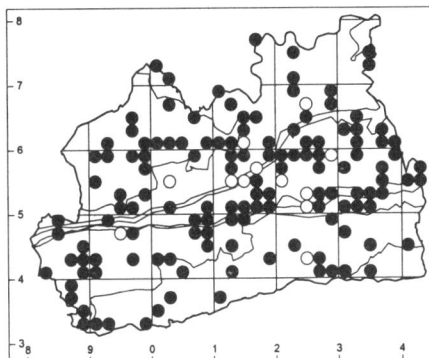

Volucella inanis (Linnaeus, 1758)   PLATE 16

Nationally Scarce
Number of records: 224
Surrey Status: Common
Flight times: June – September
Peak: August

This is a large wasp mimic whose larvae are ectoparasites of the larvae of social wasps. Although scarce on a national scale, *V. inanis* is widespread in south-east England and this is reflected by its status in Surrey. Because of its frequency, records of this species are not listed.

Flower visits: bramble, wild angelica, hogweed, upright hedge-parsley, wild marjoram, water mint, butterfly-bush, devil's-bit scabious, creeping thistle, Canadian goldenrod, yarrow, common ragwort

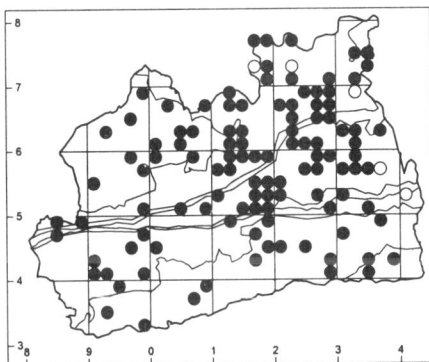

Volucella inflata (Fabricius, 1794)

Nationally Scarce
Number of records: 53
Surrey Status: Local
Flight times: June – September
Peak: July

Although principally a woodland hoverfly, *V. inflata* is sometimes found in localities with no obvious woodland. Its principal strongholds are the beech woods of the North Downs and the oak woods of the Low Weald. Until recently, this hoverfly was thought to be associated with the workings of large longhorn beetles and sap runs, which in turn are principally associated with older trees and dead timber. On a meeting of the SLENHS at Bookham Common in June 1952, "several dozen" of this species were observed feeding on a sap run on oak which was almost certainly attacked by goat moth (Richardson, 1954). This suggests an association with goat moth which is confirmed by Rotheray (1993). Don Tagg also reports seeing five individuals at a sap run on oak at Thundry Meadows. As this species appears to be far more widespread than goat moth (Collins, 1997), it is likely that the range of sap runs utilised is more diverse (confirmed by Graham Rotheray in 1997 – *pers. comm.*).

Conservation: Retention of trees infested with goat moth and of large trees in actively managed woodland is essential.

Flower visits: hogweed, elder (LP, 1950a)

RECORDS: **Thundry Meadows** SU8944 (no date, DT); **Thursley Common** SU9041 (1991, RF); **Thursley** SU9040 (7.1988, JSD); **Elstead** SU9142 (1993, RF); **Hog's Back** SU9248 (28.6.1987, RKAM); near **Willey Green** SU9450 (11.7.1987, RKAM); **Hambledon Hurst/Cuckoo Hill** SU9637 (20.6.1993, RKAM); **Prestwick Copse** SU9735 (12.7.1987, RDH); **Tugley Wood** SU9833 (1990 - 1994, JSD); **Botany Bay** SU9834 (14.7.1991, JRD); **White Beech** SU9835 (19.6.1988, RKAM); **Hoe Stream** SU9956 (18.8.1984, AJH); **Hambledon Common** (no date, LP); **Hog Wood** TQ0032 (7.7.1965, SFI); **Merrow Common** TQ0251 (5.7.1987, RKAM), (8.7.1988, GAC); **Wisley RHS Gardens** TQ0658 (7.7.1988, AJH); **Lower Canfold Wood** TQ0738 (22.7.1995, RKAM); **Hackhurst Downs** TQ0948 (21.6.1986, RKAM/GAC); **Greystones** TQ0951 (22.7.1995, HCE); **The Sheepleas** (9.6.1968, COH), TQ0851 (1982, 11.8.1991, AJH), TQ0951 (1966 - 1969, AES); **Effingham** TQ1056 (11.6.1994, RKAM); **Horsley** (6.1929, GEN); **Oaken Grove** TQ1149 (5.7.1987, PJH); **White Downs** TQ1049 (no date, JSD), TQ1249 (9.6.1985, AJH), (17.7.1985, GAC), (1992, JSD); **Leith Hill** TQ1342 (5.7.1988, DAS); **Westcott Downs** TQ1349 (16.7.1986, GAC); **Great Bookham Common** (22.6.1952, SW); **Great Oaks** TQ1562 (26.6.1988, RKAM); **Oxshott Heath** TQ1562 (20.6.1993, GAC); **Box Hill** TQ1751 (20.6.1977, SBRC), (29.6.1980, AJH), TQ1752 (2.7.1989, RKAM), TQ1851 (25.7.1992, RKAM), (1991 - 1992, JSD/RF); **Headley** (11.6.1952, LP); **Ashtead Common** (22.6.1964, SFI), TQ1859

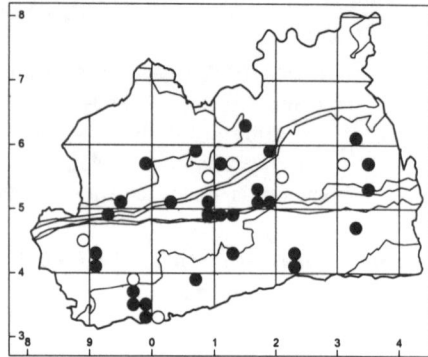

(23.8.1987, RKAM); **Glovers Wood** TQ2240 (2.7.1993, GAC); **Stanhill Court** TQ2342 (13.7.1986, RDH); **Harewoods** TQ3247 (6.8.97, KNAA); **Farthing Downs** TQ3057 (11.6.1964, SFI); **Riddlesdown** TQ3260 (31.7.1985, GAC); **Pilgrim Fort** TQ3453 (14.7.1986, GAC); **Caterham Happy Valley** TQ3456 (9.7.1987, GAC), (1.9.1985, RKAM/GAC).

Volucella pellucens (Linnaeus, 1758)   PLATE 16

Number of records: 500
Surrey Status: Ubiquitous
Flight times: May – September
Peak: July

This common and readily identifiable hoverfly is often seen hovering in woodland rides. It can be identified by the large pale windows on the abdomen which are obvious from below. There is a useful account of this species by Nixon (1934), who describes it visiting the nests of *Vespula germanica* and *V. vulgaris* at Herne Hill, with an indication that oviposition occurred in the nest of *V. germanica*.

Flower visits: bramble, dog rose, firethorn, alder buckthorn, ground-elder, hogweed, wild marjoram, water mint, butterfly-bush, devil's-bit scabious (LP, 1941), creeping thistle, common knapweed, ragwort spp.

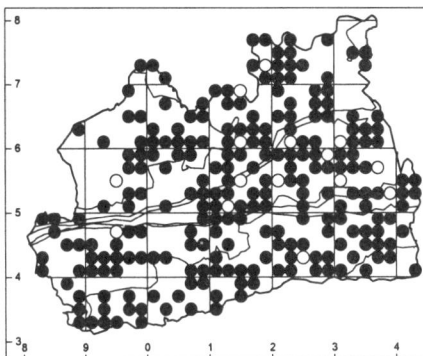

Volucella zonaria (Poda, 1761)   PLATE 16

Nationally Scarce
Number of records: 122
Surrey Status: Local
Flight times: June – October
Peak: August

Until the mid 1940s this spectacular hornet mimic, which is Britain's largest hoverfly, was known as a rare migrant to the south coast (Goffe, 1945), but at this time numbers of inland records rose rapidly, in particular around London, suggesting that a dramatic extension of range had occurred. The first Surrey record that I have traced is for Wimbledon Common on 27.8.1945 (Riley, 1946), and over subsequent years a plethora of published records suggested that *V. zonaria* had become established in the London area (e.g. Eagles, 1946, 1948, 1950; LP, 1949; Whicher, 1948; Woodcock, 1956). Published records subsequently declined as *V. zonaria* became an established part of the British fauna. Even so, it is apparent that its distribution and frequency were patchy; Danks (1963), writing from Surbiton, states that *V. zonaria* was "extremely

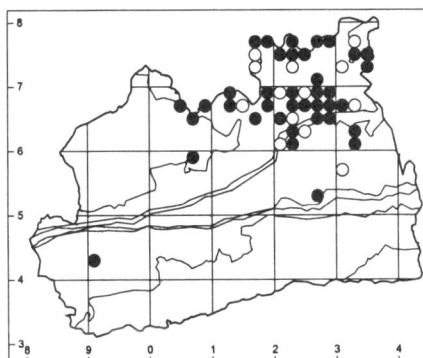

common and appears more abundantly each year", whilst Hughs (1964), writing in response to Danks, records having seen *V. zonaria* just once in the Croydon area. Even today, *V. zonaria* occurs in varying numbers each year and has not expanded its range far from urban London, with the exception of single records from Wisley, Gatton Park and Elstead. It is likely that *V. zonaria* favours the micro-climate of urban London and this may mean that it will spread further as climatic change takes a hold. This species is well known for entering buildings (Stubbs & Falk, 1983) but has also been recorded "flying around a light" at Richmond (Whicher, 1948) and was attracted to a lighted window in Kew (Ranger, 1955). Given the numbers of records of this species, individual records are not listed, but a summary of records is given in Table 12.

Flower visits: bramble, ivy (LP, 1949), wild angelica, hogweed, butterfly-bush, garden privet, *Hebe* sp., devil's-bit scabious, creeping thistle

1945-49	13
1950-54	9
1955-59	2
1960-64	6
1965-69	1
1970-74	1
1975-79	3
1980-84	11
1985	14
1986	5
1987	9
1988	1
1989	10
1990	3
1991	10
1992	0
1993	7
1994	5
1995	5
1996	2
1997	7

Figure 15. **Records of *Volucella zonaria* since 1945**

Xylotini

These are the major component of our saproxylic or dead wood fauna, with larvae inhabiting rot-holes, decaying roots, decaying sappy bark, and sawdust. One species, *Tropidia scita*, is associated with common reed and bulrush, and another, *Syritta pipiens*, is associated with decaying vegetation. Many of these species are rather uncommon, partly because their habitat is limited, but also because the habits of some make them very hard to find. Given the difficulty of finding some adults, there is scope for detailed survey of larval distribution. Members of this tribe are useful indicators of woodland and parkland habitat and are vulnerable to modern arboricultural techniques.

Genus *Brachypalpoides*

Brachypalpoides lentus (Meigen, 1822)

Number of records: 48
Surrey Status: Local
Flight times: May – September
Peak: June

This is a large and highly distinctive woodland species whose distribution reflects the major woodlands of the Chalk and the Low Weald. *B. lentus* is occasionally found at flowers such as buttercups. The larvae are associated with rot-holes and decaying roots (Rotheray, 1993), and adults which are frequently found flying around oak and beech trees at ground level may be looking for oviposition sites.

Flower visits: buttercup spp.

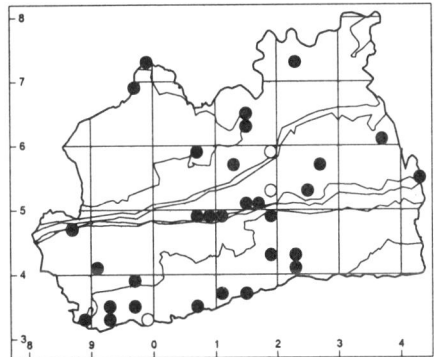

Genus *Brachypalpus*

Brachypalpus laphriformis (Fallén, 1816)   PLATE 6

Nationally Scarce

Number of records: 14

Surrey Status: Rare

Flight times: April – June

Peak: May

I have recorded this spectacular fly at hawthorn and apple blossom at Box Hill, but elsewhere have found it resting on fallen timber. Although known to breed in rot-holes in oaks, *B. laphriformis* has been found in a variety of sites such as Thursley and Wisley Commons which are not known for their tree cover. F.C. Adams (1899), discussing his experience of this species in the New Forest, described its "resemblance both on the wing and at rest to some of the Andrenidae" to be so good that he nearly overlooked it. I too have noted the same resemblance, but thought it a close fit to a species of *Osmia*. Perhaps, therefore, this is a species which should be sought by Hymenopterists.

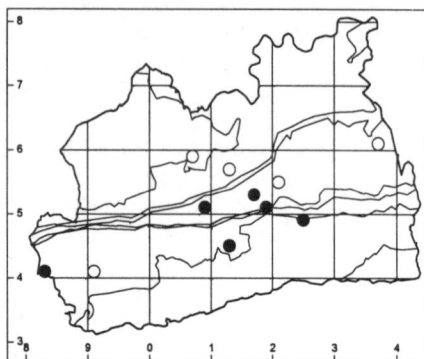

Conservation: Oaks with wet rot-holes are important for this and many other species of deadwood hoverflies; they should be retained in all circumstances.

Flower visits: apple, hawthorn

RECORDS: **Bealeswood Common** SU8240 (9.5.1987, RKAM); **Thursley Common** SU9040 (28.5.1966, AES/COH); **Weybridge** (2.6.1940, GEN); **Wisley** (28.4.1921, GFW); **Mountain Wood** TQ0950 (23.5.1993, RKAM); **Friday Street** TQ1245 (26.5.1995, GAC); **Great Bookham Common** TQ1256 (24.5.1953, PWC); **Box Hill** (10.6.1939, CNH), TQ1752 (27.5.1986, RKAM), TQ1851 (13.6.1986, RKAM); **Headley** (11.6.1965, LP); **Reigate** TQ2549 (30.4.1995, RKAM); **Selsdon** (23.5.1934, LP).

Genus *Caliprobola*

Caliprobola speciosa (Rossi, 1790)   PLATE 4

Red Data Book 1
Number of records: 2
Surrey Status: Rare
Flight times: June

Perhaps the most remarkable species recorded in Surrey is *C. speciosa* which was reported from Fairmile Common by Don Tagg, but is not backed by a voucher specimen. I have accepted this record on the basis of the recorder's familiarity with this species. This species has for some time been confined to the New Forest and Windsor Great Park, and any Surrey records might be expected from Virginia Water, although there is an old specimen from "Weybridge" in the Hope Collection at Oxford. I have re-visited Fairmile Common on a number of occasions in an attempt to verify this record. The habitat looks suitable, with a number of standing dead beech trunks in an open clearing on sandy soil, and it is possible that *C. speciosa* has become established as a result of damage from the 1987 storm which felled many beeches.

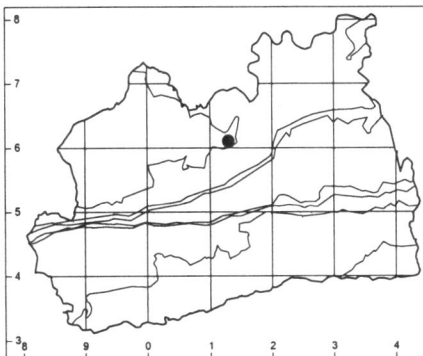

Conservation: Larvae of *C. speciosa* inhabit rotting beech roots (Rotheray, 1993) and in parkland situations this species could be vulnerable to the practice of grinding out tree stumps. Given our limited knowledge of its presence in Surrey, further survey is essential, both in the Fairmile Common area and also at Virginia Water. Other localities with old beeches on sandy soils in west Surrey should also be identified and surveyed for this and other saproxylic species.

RECORDS: **Weybridge** (no data - in Hope Collection, Oxford); **Fairmile Common** (20.6.1995, DT).

Genus *Chalcosyrphus*

Chalcosyrphus nemorum (Fabricius, 1805)

Number of records: 95

Surrey Status: Local

Flight times: April – August

Peak: June

Although *C. nemorum* is widespread, it is rarely met with in numbers and is principally associated with heathland, parts of the Low Weald and the Chalk. It is principally a woodland species and has often been found near dead or moribund beech, but is equally at home on soggy timber in wet woodland. The larvae are reported from under bark of trunks and branches in moist or wet conditions (Rotheray, 1993).

Flower visits: marsh marigold, buttercup spp., lesser spearwort, wild basil

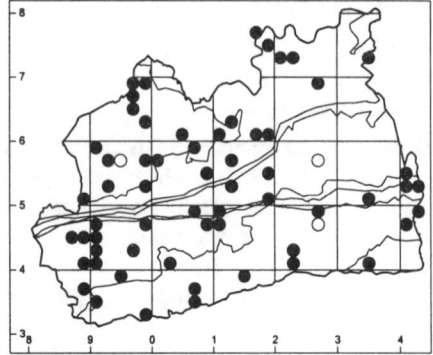

Genus *Criorhina*

This is a group of highly realistic bumblebee mimics. They generally fly early in the year, and although all four do occasionally occur together it is more usual for *C. ranunculi* to have disappeared by the time the others are on the scene. The larvae of all four species breed in decaying roots of trees and stumps, often appearing deep underground, but will also breed in rot-holes on occasions (G. Rotheray, *pers. comm.*).

Criorhina asilica (Fallén, 1816)   PLATE 4

Nationally Scarce

Number of records: 24

Surrey Status: Scarce

Flight times: May – July

Peak: May

There have been years such as 1987 when I have regarded *C. asilica* as not uncommon, and certainly when it does occur it can do so in numbers. My general impression is, however, that numbers fluctuate considerably. Adults are most commonly found at hawthorn blossom in woodland, but there are some surprising records from isolated trees on open heathland such as at Chobham Common.

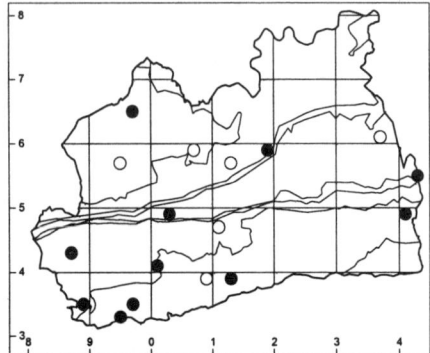

The majority of records are from south-west Surrey, especially the Chiddingfold woods. On one occasion, I caught a pair *in copula* on a sunlit sycamore leaf, having detected them from the loud buzzing that emanated from a leaf above my head.

Conservation: This hoverfly, like others in the genus, is associated with dead and decaying timber and will be best served by retention of trees which contain a large proportion of dead and decaying wood, and of moribund trees. Old stumps should also be retained and not subjected to stump grinding.

Flower visits: hawthorn, alder buckthorn, ramsons

RECORDS: **Tilford Reeds** SU8643 (28.5.1988, RKAM); **Hindhead Common** SU8934 (28.6.1987, GAC); **Killinghurst** SU9433 (25.5.1987, RKAM); **Eighty Acre Copse, Chiddingfold** SU9734 (25.5.1987, RKAM/GAC); **Chobham Longcross** SU9765 (7.6.1986, GAC); **Brookwood** (31.5.1953, LP); Woodland near **Scotslands Farm** TQ0040 (15.5.1993, RKAM); **St Martha's Hill** TQ0348 (16.5.1987, RKAM); **Wisley** TQ0658 (1966, COH); **Cranleigh** TQ0839 (4.6.1979, IFGM), (1.7.1965, SFI); **Wotton Woods** TQ1147 (1967, AES); **Chapel Copse** TQ1238 (16.5.1987, RKAM/GAC); **Great Bookham Common** (8.5.1949, LP); **Ashtead Common** TQ1859 (11.6.1986, GAC); **Selsdon** TQ3661 (14.5.1961, RLC); **Staffhurst Wood** TQ4148 (31.5.1987, RKAM/GAC); **Titsey Wood** TQ4254 (31.5.1987, RKAM/GAC).

Criorhina berberina (Fabricius, 1805)

Number of records: 66
Surrey Status: Local
Flight times: April – August
Peak: May

This widespread woodland hoverfly is clearly most common in the woods of the North Downs and on the Low Weald. It has a relatively long flight period in comparison with others of the genus. It has two colour forms; the typical form is bicoloured, tawny and black, while var. *oxyacanthae* is a uniform sandy colour. Detailed recording of the two forms has not been consistent over the survey period, so the ratio of the typical form over var. *oxyacanthae* cannot be determined.

Flower visits: raspberry, hawthorn

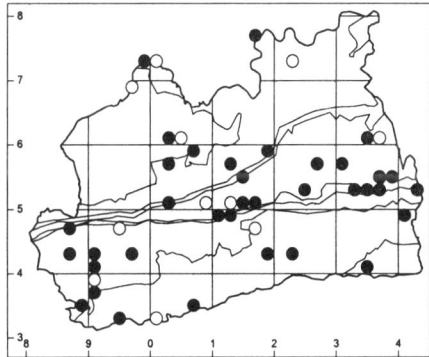

Criorhina floccosa (Meigen, 1822)   PLATE 4

Number of records: 46
Surrey Status: Local
Flight times: April – June
Peak: May

This is a widespread hoverfly which is often seen at hawthorn blossom and sunning itself on sycamore leaves. *C. floccosa* has been found in a variety of more open locations and is as much an ancient hedgerow species as it is an inhabitant of woodland. I have observed numerous females of this species flying around the base of a semi-hollowed ash tree which was a very likely breeding place at Stane Street (TQ1956) on 6.5.1990. It is likely that populations may continue for many years in a single old tree.

Flower visits: hawthorn

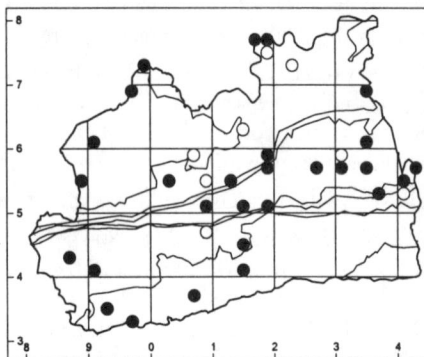

Criorhina ranunculi (Panzer, 1804)

Nationally Scarce
Number of records: 36
Surrey Status: Scarce
Flight times: March – June
Peak: April

Rotheray (1993) reports larvae in beech roots, but I have concluded that they are also associated with rotting timber in the base and roots of birch and possibly other trees, having seen females flying round such locations on a number of occasions (confirmed by Graham Rotheray, *pers. comm.*, as this text went to press). This would help to explain the relatively high frequency of *C. ranunculi* on heathlands where the dominant tree is often birch. There is also a record from Chilworth of a pair *in copula* in a cavity beneath the roots of an elm tree (Blair, 1945) which might indicate that an even wider range of rotting timber is utilised. Where it occurs, *C. ranunculi* can often be abundant and is best searched for at bird cherry and blackthorn blossom, and at sallow catkins; it can also be found resting on sunlit leaves of cherry laurel and rhododendron (Fry, 1997).

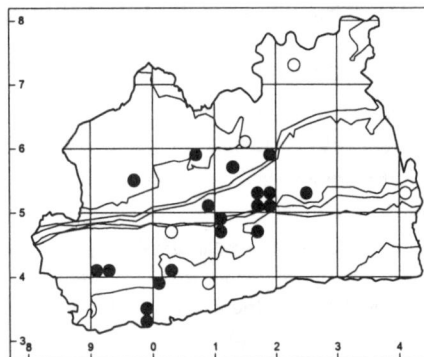

Conservation: This is one of a group of hoverflies which inhabit dead and decaying timber and is likely to occur in semi-moribund trees or trees with some internal decay; these should be retained under all circumstances. On heathland where scrub clearance is necessary, really large and ancient birches should be left wherever possible, and some future replacements retained in suitable locations. Opening up the canopy around such trees may be detrimental because the humidity regime may change and the decaying timber could dry out.

Flower visits: blackthorn, bird cherry

RECORDS: **Thursley** SU9040 (1970 - 1992, JSD); **Elstead - Thursley area** (16.3.1995, 5.5.1996, 7.5.1996, 14.5.1996, RF); **Oxted Green** SU9340 (28.3.1993, JRD); **Witley Common** SU9340 (21.3.1993, JRD), (31.3.1989, RKAM); **Worplesden** SU9654 (1.4.1989, RKAM/PLTB/GAC); **Tugley Wood** SU9833 (31.3.1989, RKAM); **White Beech** SU9835 (3.4.1988, RKAM); **Hascombe Hill** TQ0038 (22.4.1992, JSD); **Run Common** TQ0341 (18.4.1987, GAC); **Chilworth** (12.5.1934, [Blair, 1945]); **Wisley RHS Gardens** TQ0658 (1982, AJH); **Cranleigh Woods** TQ0839 (4.6.1979, IFGM/AES); **Mountain Wood** TQ0951 (25.4.1987, RKAM); **Wotton Woods** TQ1147 (3.4.1988, RKAM); **White Downs** TQ1148 (16.5.1987, AJH); **Great Bookham Common** (26.3.1948, COH), (9.5.1954, LP), (13.4.1982, MOH); **Oxshott Heath** (2.5.1940, LP); **Holmwood** TQ1746 (31.3.1989, RKAM); **Mickleham Downs** TQ1753 (4.4.1995, MSP); **Box Hill** TQ1751 (28.3.1988, RKAM), TQ1851 (1991 - 1992, JSD/RF), TQ1852 (13.3.1993, RKAM); **Ashtead Common** TQ1859 (26.4.1987, RKAM/GAC); **Wimbledon Common** (18.4.1949, AWJ); **Margery Wood** TQ2452 (30.5.1987, GAC); **Limpsfield Common** (27.3.1945, LP).

Genus *Pocota*

Pocota personata (Harris, 1780)   PLATE 5

Red Data Book 2

Number of records: 2

Surrey Status: Rare

Flight times: May – June

This species is listed on the long list of the UK Biodiversity Action Plan (DoE, 1995). There are just two old records from sites close to Richmond Park which appears to be the most appropriate area for further searches for *P. personata*. The possibility of *P. personata* turning up elsewhere in Surrey, where ancient beeches with wet rot-holes occur, should not be ruled out and such areas should be managed with species such as this in mind. Whilst there is the possibility that this species has been lost to Surrey, its elusive habits make it difficult to record and survey of rot-holes may be a useful way of establishing its true distribution.

Conservation: This is one of the suite of species whose larvae inhabit wet rot-holes in very mature trees. Arboricultural practices such as draining rot-holes and felling or trimming trees with such rot-holes should be resisted.

RECORDS: **Palewell Common** TQ2074 (20.5.1939, AML); **Putney Heath** TQ2374 "on oak leaf, clearing in scrub woodland" (2.6.1951, AWJ).

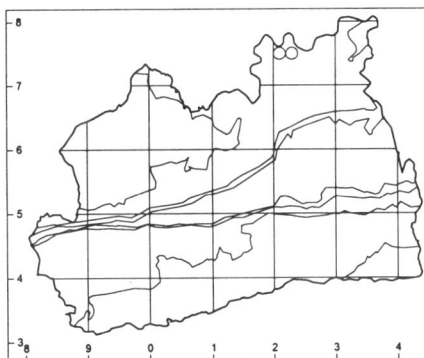

Genus *Syritta*

Syritta pipiens (Linnaeus, 1758)

Number of records: 1090
Surrey Status: Ubiquitous
Flight times: March – October
Peak: August

This is a very common hoverfly whose larval stages are associated with decaying vegetation such as in compost heaps. There is one record of this species as a prey item of the fly *Empis tessellata* (Diptera, Empididae) (LP, 1968). Six males and 10 females were taken in an insectocutor in a stables in Richmond Park in 1994, numbers which are sufficiently high as to suggest that it might be breeding in stable bedding.

Flower visits: Buttercup spp., lesser spearwort, goosefoot spp., greater stitchwort (LP, 1950a), lesser stitchwort, redshank, perforate St John's-wort, white bryony, sallow (LP, 1950a), hedge mustard, water-cress, perennial wall-rocket, white mignonette, heather, *Erica* spp., meadowsweet, bramble, tormentil, creeping cinquefoil, firethorn, hawthorn, melilot spp., ivy, wood spurge (LP, 1957), cow parsley, burnet-saxifrage, ground-elder, hemlock water-dropwort, wild angelica, wild parsnip, hogweed, upright hedge-parsley, wild carrot, field bindweed, wild marjoram, gipsywort, water mint, butterfly-bush, *Hebe* sp., eyebright, red bartsia, cleavers, devil's-bit scabious (LP, 1950a), greater burdock, creeping thistle, common knapweed, nipplewort, cat's-ear, perennial sow-thistle, smooth hawk's-beard, common fleabane, Canadian goldenrod, Michaelmas-daisy, sneezewort, yarrow, oxeye daisy, scentless mayweed, common ragwort, hoary ragwort, ragwort spp., hemp-agrimony, water-plantain

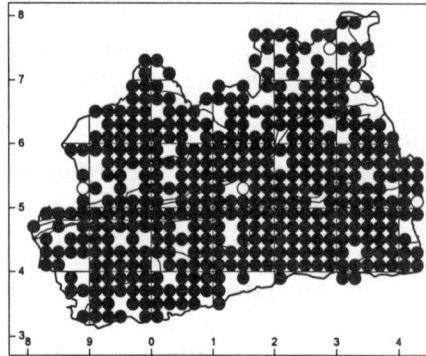

Genus *Tropidia*

Tropidia scita (Harris, 1780) PLATE 13

Number of records: 10
Surrey Status: Rare
Flight times: June – September
Peak: June

This hoverfly is usually associated with beds of common reed, *Phragmites australis*, which are not common in Surrey, but also occurs in swamps of reed sweet-grass, *Glyceria maxima*, as at Charterhouse Alderholt. The principal stronghold of the hoverfly is in north-west Surrey amongst the gravel pits of the Thorpe area, but it also seems to follow the River Wey through west Surrey. Although not a Nationally Scarce species, it is perhaps worth remarking that the conservation of this hoverfly in Surrey depends upon retention of reed beds and that such habitats are infinitely preferable to landfill and reclamation, as is often required under existing planning permissions for aggregate extraction in north-west Surrey.

RECORDS: **Thundry Meadows** SU8944 (9.1990, DT); **Thursley** SU9040 (1970 - 1992, JSD/RF); **Charterhouse Alderholt** SU9444 (8.6.1968, PJC), (30.7.1994, GAC); **Runnymede** TQ0072 (12.7.1988, RBH); **White Rose Lane NR** TQ0157 (22.6.1996, AJH); **Thorpe Meadow** TQ0270 (14.6.1986, 29.6.1986, RKAM/GAC); **Wisley RHS Gardens** TQ0659 (29.5.1997, AJH); **Chertsey Meads** TQ0665 (27.6.1987, GAC); **Esher Common** (11.7.1936, LP).

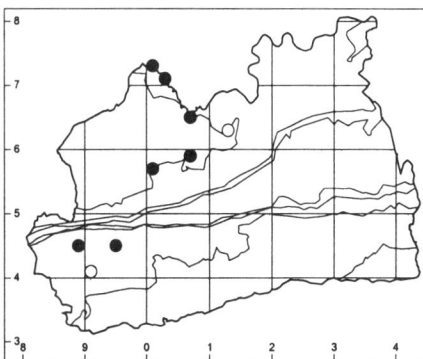

Genus *Xylota*

This is a group of medium to large species whose proportions are generally long and thin. Their larvae are chiefly associated with decaying timber, but a variety of strategies are adopted. This group rarely visits flowers, so new records of flower visits should be noted and reported.

Xylota abiens Meigen, 1822

Nationally Scarce
Number of records: 14
Surrey Status: Rare
Flight times: May – August
Peak: June

This is a scarce woodland hoverfly whose distribution matches, with few exceptions, that of the best woodlands in Surrey, such as the beech woods on the North Downs and the oak woods of the Chiddingfold complex. The larvae occur in rotting tree stumps (including conifers – G. Rotheray, *pers. comm.*). Normally *X. abiens* occurs as odd individuals, but just occasionally occurs more profusely as at Oaken Grove (TQ1149) where it was seen in large numbers at buttercup flowers on 15.6.1987 (RKAM).

Conservation: *Xylota abiens* is an indicator of high quality woodland habitat and continuity of dead wood resource; its survival depends upon sensible woodland management such as retention of cut stumps and old and moribund trees with rot-holes and other forms of decay.

Flower visits: buttercup spp., hawthorn

RECORDS: **Botany Bay** SU9734 (19.6.1988, RKAM); **Virginia Water** SU9768 (24.5.1990, RKAM); **White Beech** SU9835 (7.8.1987, RKAM); **Whitmoor Common** SU9853 (23.6.1996, AJH); **Byfleet** (9.7.1939, LP); **Wisley Common** TQ0658 (15.6.1968, 7.7.1968, PJC), (29.7.1967, AES); **Oxshott** (9.6.1895, JWY); **Oaken Grove** TQ1149 (15.6.1987, RKAM); **Great Bookham Common** TQ1256 (10.7.1986, KNAA), (25.5.1988, 6.8.1988, GAC); **Fetcham Downs** TQ1554 (4.6.1986, GAC).

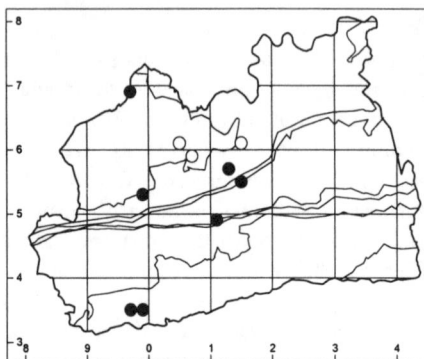

Xylota florum (Fabricius, 1805)

Nationally Scarce
Number of records: 18
Surrey Status: Rare
Flight times: June – August
Peak: July

From my encounters with *X. florum,* it appears that this hoverfly is associated with wet habitats with submerged or water-saturated timber; this may provide the clue to the larval habitat. There seems to be a dichotomy of habitats from which *X. florum* is recorded: most records are from south-west Surrey where it occurs in wet woodlands associated with the Surrey heaths, but there is also a record from alder carr in south-east Surrey.

Conservation: sites which already include boggy areas with fallen semi-submerged timber should be managed to ensure that further timber enters the system. When felling in such locations, some fallen timber of a variety of sizes should always be left in the water.

RECORDS: **Thursley** SU9040 (17.7.1991, JSD); **Elstead** SU9142 (1993, RF); **Puttenham Common** SU9145 (13.6.1987, RKAM); **Lythe Hill** SU9232 (1989, JSD); **Chobham Place Wood** SU9664 (21.6.1992, AJH); **Whitmoor Common** SU9753 (29.7.1994, RKAM); **Chobham Common** SU9765 (17.7.1989, GAC), (21.6.1992, AJH), SU9666 (13.7.1985, AJH), **Hindhead Common** SU8936 (5.8.1991, SJG); **Tuesley** SU9642 (15.7.1989, RKAM); **Oxshott Heath** (no date, GHV); **Coldharbour Common** TQ1443 (23.7.1983, KNAA); **Mosses Wood, Leith Hill** TQ1443 (25.7.1983, PSH); **Kew Gardens** TQ1676 (1.7.1987, 5.7.1988, 18.7.1988, RBH); **Hedgecourt** TQ3540 (1985, JBS).

Xylota segnis (Linnaeus, 1758)

Number of records: 529
Surrey Status: Ubiquitous
Flight times: April – October
Peaks: June, August

This common hoverfly can be found in almost any locality. The larvae develop in decaying timber and sawdust, and adults are often seen grazing over leaves covered with aphid honeydew and other debris.

Flower visits: creeping buttercup (LP, 1950a), dog rose

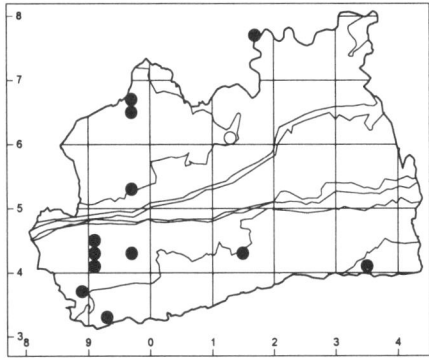

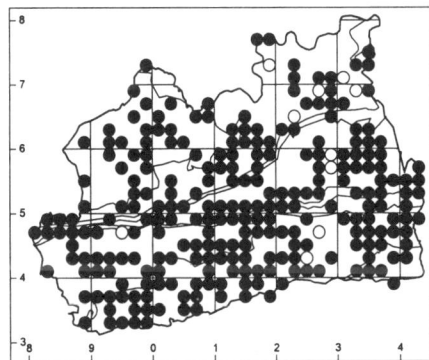

Xylota sylvarum (Linnaeus, 1758)   PLATE 4

Number of records: 166
Surrey Status: Common
Flight times: April – September
Peak: July

This is a widely distributed woodland hoverfly which is most common in central Surrey and especially on the Low Weald; it occurs mostly in old woodland, but can also be found in a wide variety of more recently wooded localities. It is occasionally seen at flowers. The larvae are known to live in decaying roots of beech and fir (Rotheray, 1993).

Flower visits: ground-elder, hemlock water-dropwort

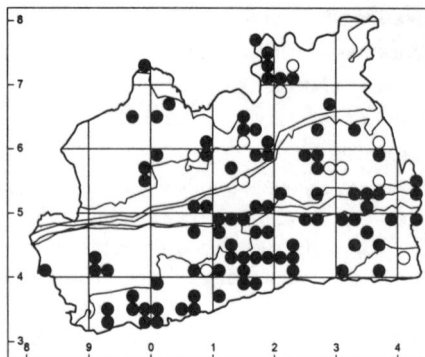

Xylota tarda Meigen, 1822

Nationally Scarce
Number of records: 4
Surrey Status: Rare
Flight times: July – August

This hoverfly is something of an enigma as there are so few records. The larvae are reported from sap runs on aspen (Rotheray, 1993) and the adults are generally found in the vicinity of aspen, but may also be associated with other poplars. Specimens of *X. segnis* should be carefully examined to ensure that *X. tarda* is not overlooked, although I have examined a good many without success. The specimen from Kew is the first for the London area since 1934 (Plant, 1986).

RECORDS: **Thursley Common** SU9041 (27.7.1966, PJC/AES); **Thursley** SU9040 (1988, JSD); **Kew Gardens** TQ1676 (18.8.1988, RBH).

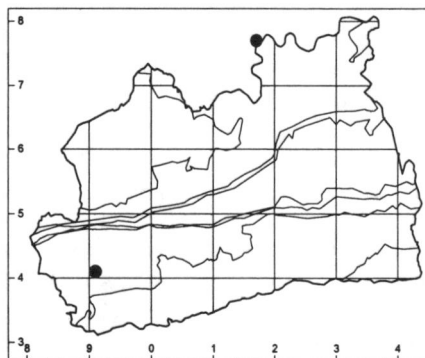

Xylota xanthocnema Collin, 1939

Nationally Scarce
Number of records: 11
Surrey Status: Rare
Flight times: June – August
Peak: July

Many records are from old woodland with large ancient trees, and parkland with mature trees. The larvae are known to inhabit rot-holes in a variety of trees, and adults are usually found on sunlit leaves. This very uncommon hoverfly is perhaps under-recorded because it is so similar to *X. sylvarum*; small gold-tailed *Xylota* should be examined carefully to ensure that they are not *X. xanthocnema*.

Conservation: Old trees with rot-holes should be retained and practices such as filling rot-holes with concrete should be avoided.

RECORDS: **Wisley RHS Gardens** TQ0658 (22.6.1981, AJH); **Hackhurst Downs** (2.7.1966, COH); **Oxshott** (26.6.1954, KMG); **Redlands Wood** TQ1545 (5.6.1993, RKAM); **Box Hill** TQ1752 (30.6.1981, AH), (2.7.1989, RKAM), TQ1852 (19.6.1988, AJH); **Nonsuch Park** TQ2363 (27.6.1984, VH); **Margery Wood/Colley Hill** TQ2452 (22.6.1953, FMS); **Earlswood Common** TQ2747 (1.8.1957, JHC); **Chelsham Wood** TQ3759 (26.7.1939, RLC).

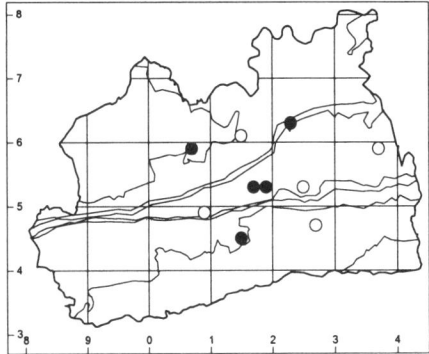

MICRODONTINAE

This is an unusual group of flies whose larvae are predators on ant grubs. The larvae are highly distinctive and shaped to protect against ant attack. There are three British species, two of which are well established in Surrey, whilst the third, *M. mutabilis*, is mainly western and northern in its distribution.

Microdon analis (Macquart, 1842) = *Microdon eggeri* Mik, 1897   PLATE 11

Nationally Scarce

Number of records: 39

Surrey Status: Scarce

Flight times: May – June

Peak: June

The best way to record this fly is to search for its larval stages which are predators of ant larvae, especially those of *Lasius niger*, on heathland. Ant nests in old pine stumps appear to be favoured as is shown by Champion (Edwards, 1923) who found a number of *Microdon* pupae under pine bark at Woking.

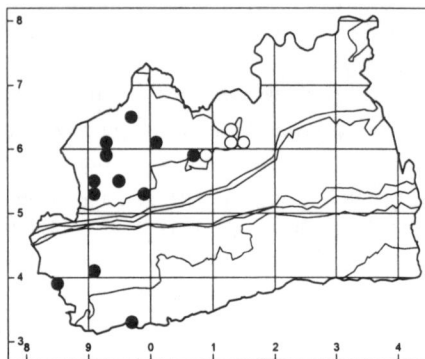

This may simply be because such stumps are easily located by the recorder, but it may help to explain why this species occurs on the southern heaths and in central Scotland (Ball & Morris, *in press*) and could make it a possibility that Scots pine is actually native to southern England. This is essentially a heathland species and one which shows some evidence of having declined. Perhaps a more surprising discovery is that of *M. analis* in the Chiddingfold woods where conifer stumps resulting from the 1987 storm seem to provide an ideal habitat for this species; an old record from Dunsfold (Wakely, 1955) suggests that *M. analis* has been established within this area for some time, however. Survey of other areas of clear-fell might therefore be worthwhile. The first British record was of a male caught at Oxshott by a Mr Grant (Verrall, 1901) and this site continued to yield records at least until the 1950's, but the lack of modern records from both Esher and Oxshott Commons suggests that *M. analis* may have been lost from these sites. It is possible that recent clearance of pines on these sites will improve the situation and, if *M. analis* survives, it should increase in numbers. Alternatively, there is the possibility that, as part of a Biodiversity Action Plan, it could be reintroduced from other parts of Surrey to re-establish its former range.

Conservation: Reversing the trend of heathland coniferisation and bracken invasion must be a priority. Where clearance of major trees is undertaken, there is an excellent opportunity for enhancing ant populations by leaving tall stumps. It would also be worth leaving large timbers in open heathland.

RECORDS: **Churt Flashes** SU8539 (6.1994, JSD); **Thursley Common** (5.1994, DT), SU9140 (1992, JSD), SU9040 (4.6.1967, AES/PJC); **Thursley** SU9040 (1970 - 1992,

JSD); **Spur Hill** SU9054 (puparium, 9.4.1989, RKAM/PLTB/GAC) (puparium,
6.5.1989, RKAM); **Wyke Common** SU9152 (puparium, 4.4.1998, GAC); **West End
Common** SU9359 (29.5.1994, RKAM); **Brentmoor Heath** SU9361 (puparium,
9.4.1989, GAC), (10.5.1992, RKAM); **Pirbright Common** SU9555 (larvae, 6.10.1991,
not reared, GAC); **Fisherlane Woods** SU9732 (5.1993, JSD); **Chiddingfold** (5.1993,
DT); **Chobham Common** SU9765 (14.6.1987, 31.5.1992, AJH), (15.6.1988, GAC);
Whitmoor Common SU9853 (11.6.1995, AJH); **Horsell Common** TQ0161 (21.4.1994,
JP), (1.6.1968, KMG); **Byfleet** (23.6.1966, LP); **Woking** (5.1923, CGC); **Dunsfold** (pre-
1954, SW); **Wisley Common** TQ0658 (24.5.1959, AES), (30.5.1968, COH), (11.6.1986,
AJH), (15.6.1986, RKAM), (6.6.1988, puparium, 3.4.1998, GAC); **Barfold Copse** (post-
1960, AES); **Esher Common** (1.6.1947, LP), (no date, RLC); **Oxshott Heath**
(9.5.1895, GHV), (24.6.1951, LP), (no date, RLC), (17.5.1954, SW).

Microdon devius (Linnaeus, 1761)   PLATE 8

Red Data Book 2
Number of records: 27
Surrey Status: Rare
Flight times: May – June
Peak: June

This is one of a small number of hoverflies
that are closely associated with chalk
downland in Surrey, and is also one which is
seriously threatened by scrub invasion of
remaining downland. This is significant
because Surrey is probably the national
stronghold for this species (Ball & Morris, *in
press*). It is especially well known from Hackhurst Downs from which there are a number
of published records (e.g. Uffen, 1964). At the time of writing, *M. devius* retains a tenuous
foothold at most of its former localities, although recent scrub clearance at sites such as
White Downs will have been highly beneficial. The flight period of this fly is quite short
and the adults are possibly difficult to spot because they do not move about much. I have
never observed flower visiting, but Wakely (1955) reports a specimen resting on a flower of
oxeye daisy. Adults are usually found near live nests of the ant *Lasius flavus,* the hill-
forming ant that inhabits dry grasslands, whose larval stages are the prey of *M. devius*
larvae (Rotheray, 1993).

Conservation: Given that this species is listed on the long list of the UK Biodiversity
Action Plan (DoE, 1995), a survey dedicated to recording the current status of this fly
might provide a useful insight into locations where efforts to reinstate downland should
concentrate and would provide the foundation for an action plan for this species. Last time
I visited Hackhurst Downs, they were heavily scrubbed up and may become unsuitable for
this species in the near future. Grassland management is the key to this species' survival,
but as it is possible that some scrub is necessary, an extensive scrub edge should be maintained
as a precaution. The only site I know of that no longer appears to support this hoverfly is

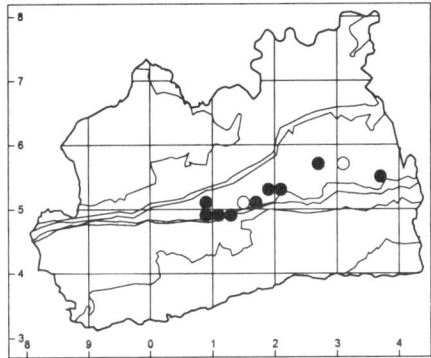

Farthing Downs, and I suspect this could be due to the introduction of a mowing regime which destroyed the ant hills in which its larvae lived.

RECORDS: **The Sheepleas** TQ0851 (29.6.1986, AJH), TQ0951 (9.6.1968, PJC), (1966 - 1969, AES); **Hackhurst Downs** TQ0948 (20.5.1948, COH), (1.6.1963, RWJU), (1970, AES), (1970, LKW), (21.6.1986, RKAM/GAC), (29.6.1986, PJH); **White Downs** (6.1975, DT), TQ1148 (15.6.1987, RKAM); **Westcott Downs** TQ1249 (28.6.1986, AJH); **Ranmore** (pre-1954, WHS); **Box Hill** TQ1751 (21.6.1986, RKAM); **Headley Warren** TQ1853 (15.6.1996, GAC); **Dawcombe** TQ2152 (7.6.1988, GAC); **Chipstead Valley** TQ2657 (12.6.1994, GAC); **Farthing Downs** TQ3057 (5.6.1959, 18.6.1959, SFI); **South Hawke** TQ3754 (1971, AES), (19.5.1990, RKAM); **Oxted** (14.6.1917, TWK).

Microdon mutabilis (Linnaeus, 1758)

Nationally Scarce
Number of records: 8
Surrey Status: Rare
Flight times: May – July
Peak: June

Most records suggest that *M. mutabilis* is associated with wet heathland, which corresponds to its known larval association with *Myrmica ruginodis,* a red ant which occupies cooler heathland and grasslands. *M. mutabilis* is most frequently found by sweeping and, although it may have been overlooked, past records suggest that it is genuinely scarce in Surrey.

Conservation: Wet heathland and damp grassland should be maintained by appropriate management.

RECORDS: **Thursley Common** SU9040 (2.6.1968, AES); **Thursley** SU9040 (1970 - 1992, RF); **West End Common** SU9360 (15.6.1988, GAC); **Horsell Common** TQ0160 (9.7.1983, AJH); **Oxshott Heath** (8.6.1947, OWR); **Wisley Common** (30.5.1968, COH), TQ0658 (15.6.1986, RKAM); **Great Bookham Common** (29.5.1949, PWC).

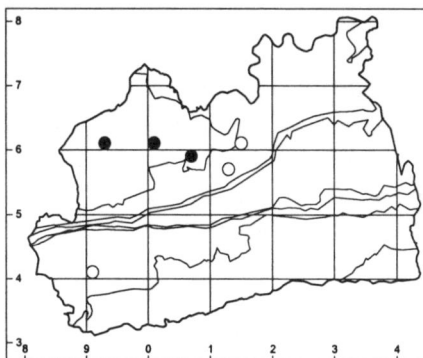

Appendices

APPENDIX 1 – Checklist of Surrey hoverflies

The nomenclature used in this text follows that of Chandler (*in press*); however the suprageneric structure used by Stubbs & Falk (1983) is retained in order that those readers comparing the species lists are not wholly confused by the changes which have been accepted in the new checklist. In order to help those readers who are not familiar with the revised nomenclature, I have indicated the change beneath the new name, as recommended by Alan Stubbs in the appendix to the reprint of Stubbs & Falk (*loc. cit.*).

* No post-1980 records	{ } Unconfirmed record only

Family SYRPHIDAE

SYRPHINAE

BACCHINI

BACCHA
elongata (Fabricius, 1775)
= *obscuripennis* Meigen, 1822

MELANOSTOMA
mellinum (Linnaeus, 1758)
scalare (Fabricius, 1794)

PLATYCHEIRUS

S. PACHYSPHYRIA
ambiguus (Fallén, 1817)

S. PLATYCHEIRUS s.s.
albimanus (Fabricius, 1781)
angustatus (Zetterstedt, 1843)
clypeatus s.s. (Meigen, 1822)
discimanus Loew, 1871
fulviventris (Macquart, 1829)
immarginatus (Zetterstedt, 1849)*
manicatus (Meigen, 1822)
occultus Goeldlin de Tiefenau,
         Maibach & Speight, 1990
peltatus (Meigen, 1822)
{*scambus* (Staeger, 1843)}
scutatus (Meigen, 1822)
sticticus (Meigen, 1822)*
tarsalis (Schummel, 1837)

S. PYROPHAENA
granditarsus (Forster, 1771)
rosarum (Fabricius, 1787)

XANTHANDRUS
comtus (Harris, 1780)

PARAGINI

PARAGUS

S. PANDASYOPTHALMUS
haemorrhous Meigen, 1822
tibialis (Fallén, 1817)

S. PARAGUS s.s.
albifrons (Fallén, 1817)*

SYRPHINI

CHRYSOTOXUM
{*arcuatum* (Linnaeus, 1758)}
bicinctum (Linnaeus, 1758)
cautum (Harris, 1776)
elegans Loew, 1841
festivum (Linnaeus, 1758)
octomaculatum Curtis, 1837
verralli Collin, 1940

DASYSYRPHUS
albostriatus (Fallén, 1817)
{*friuliensis* (van der Goot, 1960)}
pinastri (De Geer, 1776)
         = *lunulatus*: auctt., misident
tricinctus (Fallén, 1817)
venustus (Meigen, 1822)

DIDEA
fasciata Macquart, 1834
intermedia Loew, 1854

DOROS
profuges (Harris, 1780)
         = *conopseus* (Fabricius, 1775)

EPISTROPHE
diaphana (Zetterstedt, 1843)
eligans (Harris, 1780)
grossulariae (Meigen, 1822)
melanostoma (Zetterstedt, 1843)
nitidicollis (Meigen, 1822)
EPISYRPHUS
balteatus (De Geer, 1776)
ERIOZONA
erratica (Linnaeus, 1758)*
    = *Megasyrphus annulipes*
    (Zetterstedt, 1838)
EUPEODES
    = **METASYRPHUS**
corollae (Fabricius, 1794)
latifasciatus (Macquart, 1829)
latilunulatus (Collin, 1931)
luniger (Meigen, 1822)
nitens (Zetterstedt, 1843)
LEUCOZONA
S. ISCHYROSYRPHUS
glaucia (Linnaeus, 1758)
laternaria (Müller, 1776)
S. LEUCOZONA s.s.
lucorum (Linnaeus, 1758)
MELANGYNA
barbifrons (Fallén, 1817)
cincta (Fallén, 1817)
compositarum (Verrall, 1873)
labiatarum (Verrall, 1901)
lasiophthalma (Zetterstedt, 1843)
quadrimaculata (Verrall, 1873)
umbellatarum (Fabricius, 1794)
MELIGRAMMA
euchromum (Kowarz, 1885)
    = *Epistrophella euchroma*
    (Kowarz, 1885)
guttatum (Fallén, 1817)
trianguliferum (Zetterstedt, 1843)

MELISCAEVA
auricollis (Meigen, 1822)
cinctella (Zetterstedt, 1843)
PARASYRPHUS
annulatus (Zetterstedt, 1838)
lineola (Zetterstedt, 1843)
malinellus (Collin, 1952)
punctulatus (Verrall, 1873)
vittiger (Zetterstedt, 1843)
SCAEVA
pyrastri (Linnaeus, 1758)
selenitica (Meigen, 1822)
SPHAEROPHORIA
batava Goeldlin de Tiefenau, 1974
fatarum Goeldlin de Tiefenau, 1989
    = *abbreviata*: auctt., misident.
interrupta (Fabricius, 1805)
    = *menthastri*: Vockeroth, 1963,
    misident.
philanthus (Meigen, 1822)
rueppellii (Wiedemann, 1830)
scripta (Linnaeus, 1758)
taeniata (Meigen, 1822)
virgata Goeldlin de Tiefenau, 1974
SYRPHUS
ribesii (Linnaeus, 1758)
torvus Osten Sacken, 1875
vitripennis Meigen, 1822
XANTHOGRAMMA
citrofasciatum (De Geer, 1776)
pedissequum (Harris, 1776)

MILESIINAE

CALLICERINI

CALLICERA
aurata (Rossi, 1790)
    = *aenea*: auctt., misident.

CHEILOSINI

CHEILOSIA
albipila Meigen, 1838
albitarsis (Meigen 1822)
antiqua (Meigen, 1822)
barbata Loew, 1857
bergenstammi Becker, 1894
carbonaria Egger, 1860
{*cynocephala* Loew, 1840}
fraterna (Meigen, 1830)
griseiventris Loew, 1857
grossa (Fallén, 1817)
illustrata (Harris, 1780)
impressa Loew, 1840
lasiopa Kowarz, 1885
    = *honesta*: Verrall, 1901,
    misident.
latifrons (Zetterstedt, 1843)
    = *intonsa* Loew, 1857
longula (Zetterstedt, 1838)
mutabilis (Fallén, 1817)
nebulosa Verrall, 1871
nigripes (Meigen, 1822)
pagana (Meigen, 1822)
praecox (Zetterstedt, 1843)
proxima (Zetterstedt, 1843)
    = Species D & E of Stubbs & Falk, 1983
scutellata (Fallén, 1817)
semifasciata Becker, 1894
soror (Zetterstedt, 1843)
variabilis (Panzer, 1798)
velutina Loew, 1840
vernalis (Fallén, 1817)
{*vicina* (Zetterstedt, 1849)}
    = *nasutula* Becker, 1894
vulpina (Meigen, 1822)

FERDINANDEA
cuprea (Scopoli, 1763)
ruficornis (Fabricius, 1775)

PORTEVINIA
maculata (Fallén, 1817)

RHINGIA
campestris Meigen, 1822
rostrata (Linnaeus, 1758)

CHRYSOGASTRINI

BRACHYOPA
bicolor (Fallén, 1817)
insensilis Collin, 1939
pilosa Collin, 1939
scutellaris Robineau-Desvoidy, 1843

CHRYSOGASTER
cemiteriorum (Linnaeus, 1758)
    = *chalybeata* Meigen, 1822
solstitialis (Fallén, 1817)
virescens Loew, 1854

LEJOGASTER
metallina (Fabricius, 1781)
tarsata (Megerle *in* Meigen, 1822)
    = *splendida* (Meigen, 1822)*

MELANOGASTER
aerosa (Loew, 1843)
    = *macquarti*: auctt., misident.
hirtella Loew, 1843

MYOLEPTA
dubia (Fabricius, 1805)
    = *luteola* (Gmelin, 1788)

NEOASCIA

S. NEOASCIA
podagrica (Fabricius, 1775)
tenur (Harris, 1780)

S. NEOASCIELLA
geniculata (Meigen, 1822)
interrupta (Meigen, 1822)
meticulosa (Scopoli, 1763)
obliqua Coe, 1940

ORTHONEVRA
brevicornis (Loew, 1843)
geniculata (Meigen, 1830)
nobilis (Fallén, 1817)

RIPONNENSIA
splendens (Meigen, 1822)

SPHEGINA

S. SPHEGINA s.s.
clunipes (Fallén, 1816)
elegans Schummel, 1843
    = *kimakowiczi* (Strobl, 1897)
verecunda Collin, 1937

ERISTALINI

ANASIMYIA
contracta Claussen & Torp, 1980
lineata (Fabricius, 1787)
lunulata (Meigen, 1822)*
transfuga (Linnaeus, 1758)

ERISTALINUS

S. ERISTALINUS s.s.
sepulchralis (Linnaeus, 1758)

S. LATHYROPHTHALMUS
aeneus (Scopoli, 1763)*

ERISTALIS

S. EOSERISTALIS
{*abusivus* Collin, 1931}
arbustorum (Linnaeus, 1758)
horticola (De Geer, 1776)
interruptus (Poda, 1761)
        = *nemorum*: auctt., misident.
intricarius (Linnaeus, 1758)
pertinax (Scopoli, 1763)

S. ERISTALIS s.s.
tenax (Linnaeus, 1758)

HELOPHILUS
hybridus Loew, 1846
pendulus (Linnaeus, 1758)
trivittatus (Fabricius, 1805)

MALLOTA
cimbiciformis (Fallén, 1817)*

MYATHROPA
florea (Linnaeus, 1758)

PARHELOPHILUS
frutetorum (Fabricius, 1775)
versicolor (Fabricius, 1794)

MERODONTINI

EUMERUS
ornatus Meigen, 1822
strigatus (Fallén, 1817)
tuberculatus Rondani, 1857

MERODON
equestris (Fabricius, 1794)

PSILOTA
anthracina Meigen, 1822

PELECOCERINI

PELECOCERA
tricincta Meigen, 1822

PIPIZINI

HERINGIA

S. HERINGIA s.s.
heringi (Zetterstedt, 1843)

S. NEOCNEMODON
brevidens (Egger, 1865)*
latitarsis (Egger, 1865)
pubescens (Delucchi & Pschorn-
    Walcher, 1955)
vitripennis (Meigen, 1822)

PIPIZA
austriaca Meigen, 1822
bimaculata Meigen, 1822
fenestrata Meigen, 1822
lugubris (Fabricius, 1775)
luteitarsis Zetterstedt, 1843
noctiluca (Linnaeus, 1758)

PIPIZELLA
maculipennis (Meigen, 1822)*
viduata (Linnaeus, 1758)
    = *varipes* (Meigen, 1822)
virens (Fabricius, 1805)

TRICHOPSOMYIA
flavitarsis (Meigen, 1822)

TRIGLYPHUS
primus Loew, 1840

SERICOMYIINI

SERICOMYIA
lappona (Linnaeus, 1758)
silentis (Harris, 1776)

VOLUCELLINI

VOLUCELLA
bombylans (Linnaeus, 1758)
inanis (Linnaeus, 1758)
inflata (Fabricius, 1794)
pellucens (Linnaeus, 1758)
zonaria (Poda, 1761)

XYLOTINI

BRACHYPALPOIDES
lentus (Meigen, 1822)

BRACHYPALPUS
laphriformis (Fallén, 1816)

CALIPROBOLA
speciosa (Rossi, 1790)

CHALCOSYRPHUS

S. XYLOTINA
nemorum (Fabricius, 1805)

CRIORHINA
asilica (Fallén, 1816)
berberina (Fabricius, 1805)
floccosa (Meigen, 1822)
ranunculi (Panzer, 1804)

POCOTA
personata (Harris, 1780)*

SYRITTA
pipiens (Linnaeus, 1758)

TROPIDIA
scita (Harris, 1780)

XYLOTA
abiens Meigen, 1822
florum (Fabricius, 1805)
segnis (Linnaeus, 1758)
sylvarum (Linnaeus, 1758)
tarda Meigen, 1822
xanthocnema Collin, 1939

MICRODONTINAE

MICRODON
analis (Macquart, 1842)
    = *eggeri* Mik, 1897
devius (Linnaeus, 1761)
mutabilis (Linnaeus, 1758)

NUMBER OF SPECIES RECORDED

1980 to date 198

Pre - 1980 11

Unconfirmed and Doubtful 5

Total species 214

Total GB fauna .... +/- 270

APPENDIX 2 – Flowers visited by hoverflies

Records in this section comprise modern records submitted as part of the recording scheme, and literature records which are annotated accordingly. They relate solely to flower visits by hoverflies in Surrey.

Ranunculaceae

Caltha palustris, **marsh marigold (8):** *M. mellinum, P. albimanus, M. cinctella, C. antiqua, C. pagana, E. pertinax, H. brevidens* (Stubbs, 1980), *C. nemorum*

Anemone nemorosa, **wood anemone (1):** *N. podagrica*

Clematis vitalba, **traveller's-joy (2):** *S. pyrastri, R. campestris*

Ranunculus acris, **meadow buttercup (12):** *C. bicinctum, C. festivum, E. balteatus, M. auricollis, S. scripta, S. ribesii, C. albitarsis, C. fraterna, M. hirtella, A. lineata, E. sepulchralis, P. viduata*

Ranunculus repens, **creeping buttercup (31):** *M. mellinum* (LP, 1950a), *M. scalare, P. albimanus, P. rosarum* (LP, 1950a), *C. bicinctum, D. albostriatus* (LP, 1950a), *D. tricinctus* (LP, 1950a), *D. venustus* (LP, 1950a), *E. balteatus, E. luniger* (LP, 1950a), *L. lucorum, M. lasiophthalma* (LP, 1950a), *M. auricollis, S. scripta, S. ribesii* (LP, 1950a), *X. pedissequum, C. albitarsis, C. fraterna* (LP, 1950a), *C. pagana, C. variabilis, M. hirtella, N. geniculata* (LP, 1950a), *N. tenur* (LP, 1950a), *E. arbustorum, E. interruptus* (LP, 1950a), *E. pertinax, E. tenax* (LP, 1950a), *H. pendulus* (LP, 1950a), *E. tuberculatus, P. fenestrata* (LP, 1960), *X. segnis* (LP, 1950a)

Ranunculus bulbosus, **bulbous buttercup (4):** *L. lucorum, S. scripta, C. albitarsis, P. viduata*

Ranunculus **spp., buttercup spp. (55):** *M. mellinum, M. scalare, P. albimanus, P. clypeatus* agg., *P. peltatus, C. bicinctum, C. cautum, C. octomaculatum* (Miles, 1989), *D. tricinctus, D. venustus, E. eligans, E. melanostoma, E. balteatus, E. corollae, E. latifasciatus, E. luniger, L. lucorum, M. auricollis, S. taeniata, S. ribesii, S. vitripennis, C. albitarsis, C. antiqua, C. bergenstammi, C. fraterna, C. illustrata, C. lasiopa, C. nigripes, C. pagana, C. vernalis, F. cuprea, M. hirtella, L. metallina, N. obliqua, N. podagrica, O. nobilis, R. splendens, A. contracta, A. lineata, E. sepulchralis, E. arbustorum, E. interruptus, E. pertinax, E. tenax, H. pendulus, E. tuberculatus, M. equestris, P. noctiluca, P. viduata, P. virens, S. lappona, B. lentus, C. nemorum, S. pipiens, X. abiens*

Ranunculus flammula, **lesser spearwort (10):** *M. mellinum* (LP, 1950a), *P. granditarsus, C. bicinctum, E. balteatus, M. auricollis* (LP, 1950a), *C. pagana, H. pendulus* (LP, 1950a), *M. florea, C. nemorum, S. pipiens*

Ranunculus ficaria, **lesser celandine (11):** *M. scalare, P. albimanus, E. latifasciatus, L. lucorum, M. cincta, S. scripta* (LP, 1950a), *C. albitarsis, C. pagana, C. vernalis, R. campestris* (LP, 1950a), *E. pertinax* (LP, 1950a)

Ranunculus **subgen.** *Batrachium,* **water crowfoot (2):** *P. albimanus* (LP, 1950a), *E. sepulchralis* (LP, 1950a)

Papaveraceae

Papaver somniferum, **opium poppy (1):** *E. balteatus*

P. rhoeas, **common poppy (1):** *E. balteatus*

Urticaceae

Urtica dioica, **common nettle (1):** *P. scutatus*

Chenopodiaceae

Chenopodium vulvaria, **stinking goosefoot (1):** *S. rueppellii* (Iliff, 1991)

Chenopodium **spp., goosefoot spp. (5):** *P. granditarsus, P. rosarum, E. balteatus, E. pertinax, S. pipiens*

Caryophyllaceae

Stellaria media, **chickweed (3):** *P. albimanus, P. scutatus, E. luniger*

Stellaria holostea, **greater stitchwort (17):** *M. mellinum* (LP, 1950a), *M. scalare, P. albimanus, P. manicatus* (LP, 1950a), *P. tarsalis, D. albostriatus, E. balteatus, L. lucorum, S. scripta* (LP, 1950a), *S. ribesii, C. pagana, F. cuprea, E. sepulchralis, E. arbustorum* (LP, 1950a), *M. equestris, P. noctiluca* (LP, 1950a), *S. pipiens* (LP, 1950a)

Stellaria graminea, **lesser stitchwort (10):** *M. scalare, P. albimanus, C. bicinctum, E. balteatus, E. luniger, M. auricollis, S. scripta, C. pagana, E. tuberculatus, S. pipiens*

Cerastium glomeratum, **sticky mouse-ear (1):** *S. ribesii*

Myosoton aquaticum, **water chickweed (6):** *M. scalare, P. albimanus, P. scutatus, E. balteatus, C. pagana, R. campestris*

Spergularia rubra, **sand spurrey (1):** *S. scripta*

Lychnis flos-cuculi, **ragged robin (2):** *E. luniger* (LP, 1950a), *R. campestris*

Polygonaceae

Persicaria maculosa, **redshank (12):** *P. clypeatus* s.s., *P. scutatus, P. granditarsus, P. rosarum, E. balteatus, E. latifasciatus, E. luniger, S. rueppellii, S. scripta, E. arbustorum, E. pertinax, S. pipiens*

Fagopyrum esculentum, **buckwheat (1):** *E. latifasciatus*

Polygonum aviculare, **knotgrass (6):** *P. scutatus, Paragus* sp., *E.balteatus, E. luniger, M. florea, E. tuberculatus*

Fallopia japonica, **Japanese knotweed (7):** *P. scutatus, E. balteatus, C. pagana, E. arbustorum, E. pertinax, E. tenax, E. tuberculatus*

Rumex sanguineus, **wood dock (1):** *E. luniger*

Rumex obtusifolius, **broad-leaved dock (1):** *M. scalare*

Clusiaceae

Hypericum perforatum, **perforate St John's-wort (7):** *E. balteatus, E. corollae, E. luniger, S. vitripennis* (LP, 1950a), *F. cuprea, E. tenax, S. pipiens*

Hypericum maculatum, **imperforate St John's-wort (2):** *S. vitripennis, E. tenax*

Cistraceae

Helianthemum nummularium, **common rock-rose (4):** *P. haemorrhous, C. bicinctum, C. festivum, P. viduata*

Violaceae

Viola **spp., violet (1):** *R. campestris* (LP, 1950a)

Cucurbitaceae

Bryonia dioica, **white bryony (2):** *E. balteatus, S. pipiens*

Salicaceae

Salix repens, **creeping willow (1):** *C. praecox*

Salix caprea/cinerea, **sallow (12):** *P. discimanus, M. barbifrons, M. lasiophthalma* (LP, 1950a), *M. quadrimaculata, P. punctulatus, S. torvus* (LP, 1950a), *C. albipila, C. grossa, O. geniculata, E. arbustorum* (LP, 1950a), *E. tenax* (LP, 1950a), *S. pipiens* (LP, 1950a)

Brassicaceae

Sisymbrium officinale, **hedge mustard (8):** *P. albimanus* (LP, 1950a), *E. balteatus, E. luniger, M. auricollis, Sphaerophoria* sp., *S. vitripennis, R. campestris* (LP, 1950a), *S. pipiens*

Alliaria petiolata, **garlic mustard (19):** *M. scalare, P. albimanus, P. clypeatus* agg., *P. scutatus, P. tarsalis, D. venustus, E. balteatus, E. luniger, L. lucorum, M. auricollis, M. cinctella, P. punctulatus, S. clunipes, C. albitarsis, C. pagana, R. campestris, N. podagrica, E. arbustorum, E. interruptus*

Barbarea vulgaris, **winter-cress (1):** *E. balteatus*

Rorippa nasturtium-aquaticum, **water-cress (12):** *P. granditarsus, C. festivum, S. rueppellii, C. impressa, N. podagrica, A. contracta, E. sepulchralis, E. arbustorum, E. interruptus, E. pertinax, P. versicolor, S. pipiens*

Armoracia rusticana, **horse-radish (1):** *M. auricollis*

Cardamine pratensis, **cuckoo-flower (4):** *M. mellinum* (LP, 1950a), *P. albimanus* (LP, 1950a), *L. lucorum, R. campestris*

Diplotaxis tenuifolia, **perennial wall-rocket (7):** *P. scutatus, E. balteatus, S. ribesii, C. pagana, E. arbustorum, E. tenax, S. pipiens*

Brassica oleracea, **wild cabbage (2):** *E. sepulchralis, P. versicolor*

Brassica napus, **rape (1):** *E. tenax*

Sinapis arvensis, **charlock (4):** *C. bicinctum, E. balteatus, E. corollae, E. luniger*

Rapistrum rugosum, **bastard cabbage (3):** *M. mellinum, L. metallina, E. sepulchralis*

Resedaceae

Reseda luteola, **weld (2):** *C. festivum, E. balteatus*

Reseda alba, **white mignonette (1):** *S. pipiens*

Ericaceae

Rhododendron ponticum, **rhododendron (1):** *P. luteitarsis*

Daboecia cantabrica, **St Dabeoc's heath (1):** *R. campestris*

Calluna vulgaris, **heather (16):** *M. mellinum, M. scalare, P. albimanus, P. angustatus, X. comtus* (Moore, 1989), *E. balteatus, E. corollae, E. latifasciatus, E. luniger, M. auricollis, S. pyrastri, S. virgata, S. ribesii, E. pertinax, H. pendulus, S. pipiens*

Erica cinerea, **bell heather (1):** *S. silentis*

Erica **spp. (5):** *P. albimanus, E. balteatus, E. tenax, H. pendulus, S. pipiens*

Vaccinium myrtillus, **bilberry (1):** *R. campestris*

Primulaceae

Primula vulgaris, **primrose (4):** *P. albimanus* (LP, 1950a), *S. ribesii* (LP, 1950a), *S. torvus* (LP, 1950a), *R. campestris* (LP, 1950a)

Lysimachia nemorum, **yellow pimpernel (1):** *P. manicatus* (LP, 1950a)

Lysimachia vulgaris, **yellow loosestrife (1):** *E. balteatus*

Rosaceae

Filipendula vulgaris, **dropwort (1):** *E. tenax*

Filipendula ulmaria, **meadowsweet (8):** *E. balteatus, S. vitripennis, E. arbustorum, E. interruptus, E. pertinax, E. tenax, H. pendulus, S. pipiens*

Rubus idaeus, **raspberry (1):** *C. berberina*

Rubus fruticosus **agg., bramble (22):** *M. scalare, C. bicinctum, C. festivum, D. tricinctus, E. grossulariae, E. nitidicollis, E. balteatus, S. torvus, S. vitripennis, R. campestris, N. podagrica, E. arbustorum* (LP, 1950a), *E. intricarius, E. pertinax, E. tenax* (LP, 1950a), *H. pendulus, M. florea, V. bombylans, V. inanis, V. pellucens, V. zonaria, S. pipiens*

Potentilla anserina, **silverweed (11):** *P. granditarsus* (LP, 1950a), *P. rosarum* (LP, 1950a), *D. albostriatus* (LP, 1950a), *E. balteatus, C. albitarsis* (LP, 1950a), *C. fraterna* (LP, 1950a), *E. sepulchralis* (LP, 1950a), *E. arbustorum* (LP, 1950a), *E. interruptus* (LP, 1950a), *E. tenax* (LP, 1950a), *H. pendulus* (LP, 1950a)

Potentilla erecta, **tormentil (11):** *M. mellinum* (LP, 1950a), *P. albimanus, C. bicinctum* (LP, 1950a), *S. virgata, N. tenur, E. sepulchralis* (LP, 1950a), *E. pertinax* (LP, 1950a), *M. equestris* (LP, 1950a), *E. tuberculatus, P. viduata, S. pipiens*

Potentilla reptans, **creeping cinquefoil (4):** *C. pagana, H. pendulus, P. viduata, S. pipiens*

Geum urbanum, **wood avens (1):** *M. scalare*

Agrimonia eupatoria, **agrimony (1):** *C. festivum*

Rosa canina, **dog rose (6):** *E. melanostoma, S. ribesii, E. pertinax* (LP, 1950a), *H. pendulus, V. pellucens, X. segnis*

Prunus spinosa, **blackthorn (15):** *M. scalare, P. ambiguus, P. albimanus, P. discimanus, E. eligans, E. luniger* (LP, 1950a), *M. euchromum, M. auricollis, P. punctulatus, S. torvus, R. campestris* (LP, 1950a), *E. arbustorum* (LP, 1950a), *E. intricarius* (LP, 1950a), *E. pertinax, C. ranunculi*

Prunus avium, **wild cherry (4):** *M. scalare, P. albimanus, C. praecox, E. pertinax*

Prunus padus, **bird cherry (7):** *M. lasiophthalma, M. quadrimaculata, M. auricollis, P. punctulatus, C. pagana, E. tenax, C. ranunculi*

Prunus laurocerasus, **cherry laurel (5):** *D. albostriatus, E. eligans, E. luniger, M. auricollis, P. annulatus, E. pertinax*

Malus **spp., apple (2):** *H. heringi, B. laphriformis*

Sorbus aucuparia, **rowan (12):** *M. mellinum, P. albimanus, D. venustus, E. eligans, L. lucorum, M. cinctella, S. vitripennis, C. pagana, E. arbustorum, E. pertinax, H. pendulus, S. lappona*

Pyracantha coccinea, **firethorn (9):** *D. venustus, E. eligans, E. luniger, C. aurata, E. pertinax, E. tenax, M. florea, V. pellucens, S. pipiens*

Crataegus **spp., hawthorn (38):** *M. mellinum, M. scalare* (LP, 1957), *P. albimanus, D. venustus, E. eligans, E. nitidicollis, E. balteatus, E. luniger* (LP, 1957), *L. lucorum, M. euchromum* (LP, 1950a), *M. cinctella, P. punctulatus, S. scripta* (LP, 1950a), *S. ribesii* (LP, 1950a), *S. torvus* (LP, 1957), *S. vitripennis, X. citrofasciatum, C. carbonaria* (LP, 1950a), *C. vernalis* (LP, 1950a), *R. campestris, R. rostrata* (Coe, 1961), *E. arbustorum, E. horticola, E. intricarius, E. interruptus, E. pertinax, E. tenax, H. pendulus, M. florea, E. tuberculatus, P. anthracina, P. austriaca, B. laphriformis, C. asilica, C. berberina, C. floccosa, S. pipiens, X. abiens*

Fabaceae

Anthyllis vulneraria, **kidney vetch (1):** *S. scripta*

Melilotus **spp., melilot spp. (1):** *S. pipiens*

Medicago lupulina, **black medick (4):** *P. albimanus, P. clypeatus* s.s., *Paragus* spp., *S. scripta*

Trifolium pratense, **red clover (1):** *E. luniger*

Cytisus scoparius, **broom (1):** *C. virescens*

Onagraceae

Epilobium hirsutum, **great willowherb (2):** *Sphaerophoria* spp., *R. campestris*

Epilobium tetragonum, **square-stalked willowherb (4):** *P. albimanus, P. peltatus, S. taeniata, H. pendulus*

Chamerion angustifolium, **rosebay willowherb (1):** *M. scalare*

Circaea lutetiana, **enchanter's nightshade (2):** *P. scutatus, S. clunipes*

Cornaceae

Cornus sanguinea, **dogwood (8):** *M. scalare, C. cautum, R. rostrata* (Coe, 1961), *E. sepulchralis, E. tenax, H. hybridus, H. pendulus, M. florea*

Aquifoliaceae

Ilex aquifolium, **holly (1):** *M. florea*

Euphorbiaceae

Mercuralis perennis, **dog's mercury (2):** *M. scalare, E. eligans*

Euphorbia amygdaloides, **wood spurge (21):** All records from LP (1957) unless stated: *M. mellinum, P. albimanus, D. albostriatus, E. eligans, E. corollae, E. luniger, L. lucorum, M. euchromum* (LP, 1954), *S. scripta, S. ribesii, S. vitripennis, C. bergenstammi, C. pagana, C. variabilis, R. campestris, E. pertinax, M. florea, E. tuberculatus, P. noctiluca, S. pipiens, V. bombylans*

Rhamnaceae

Frangula alnus, **alder buckthorn (2):** *V. pellucens, C. asilica*

Aceraceae

Acer platanoides, **Norway maple (3):** *E. eligans, S. torvus, E. pertinax*

Acer campestre, **field maple (5):** *D. albostriatus, E. eligans, L. lucorum, M. auricollis, S. vitripennis*

Acer pseudoplatanus, **sycamore (4):** *E. eligans, M. cincta, M. cinctella, R. campestris*

Geraniaceae

Geranium robertianum, **herb robert (5):** *P. albimanus, P. tarsalis, E. balteatus, L. lucorum, R. campestris*

Balsaminaceae

Impatiens glandulifera, **Indian balsam (1):** *R. campestris*

Araliaceae

Hedera helix, **ivy (18):** *M. mellinum, M. scalare, P. albimanus, E. grossulariae, E. balteatus, M. cinctella, S. ribesii, S. torvus, S. vitripennis, C. scutellata, E. pertinax, E. tenax, H. pendulus, M. florea, S. silentis, V. zonaria (LP, 1949), S. pipiens*

Apiaceae

Chaerophyllum temulum, **rough chervil (1):** *C. soror*

Anthriscus sylvestris, **cow parsley (40):** *Baccha* spp., *M. scalare, P. albimanus, P. tarsalis, P. scutatus, C. cautum, D. albostriatus, E. diaphana, E. eligans, E. melanostoma, E. nitidicollis, E. balteatus, E. luniger, L. lucorum, M. labiatarum, M. auricollis, S. rueppellii, S. scripta, S. ribesii, S. vitripennis, C. impressa, C. lasiopa, C. pagana, C. proxima, C. variabilis, C. vulpina, R. campestris, C. solstitialis, N. podagrica, R. splendens, S. elegans, S. verecunda, E. arbustorum, E. pertinax, H. pendulus, M. florea, P. versicolor, P. noctiluca, P. viduata, S. pipiens*

Conopodium majus, **pignut (1):** *Baccha* spp.

Pimpinella saxifraga, **burnet-saxifrage (23):** *P. albimanus, C. bicinctum, C. festivum, E. diaphana, E. balteatus, E. corollae, M. umbellatarum (LP, 1966), S. pyrastri, S. scripta, S. taeniata, S. vitripennis, C. impressa, C. pagana, C. proxima, C. soror, C. cemiteriorum, C. solstitialis, O. nobilis, R. splendens, E. tenax, M. florea, P. viduata, S. pipiens*

Aegopodium podagaria, **ground-elder (20):** *E. grossulariae, E. balteatus, L. lucorum, M. labiatarum, C. illustrata, C. pagana, C. soror, C. solstitialis, O. nobilis, R. splendens, E. arbustorum, E. interruptus, E. pertinax, E. tenax, H. pendulus, M. florea, P. viduata, V. pellucens, S. pipiens, X. sylvarum*

Oenanthe crocata, **hemlock water-dropwort (20):** *P. albimanus, D. albostriatus, L. lucorum, M. labiatarum, S. ribesii, S. vitripennis, C. proxima, C. solstitialis, O. brevicornis, R. splendens, E. sepulchralis, E. arbustorum, E. interruptus, E. pertinax, E. tenax, H. pendulus, M. florea, P. versicolor, S. pipiens, X. sylvarum*

Silaum silaus, **pepper-saxifrage (1):** *E. luniger* (LP, 1950a)

Conium maculatum, **hemlock (1):** *M. florea*

Apium nodiflorum, **fool's water-cress (7):** *E. balteatus, E. corollae, C. impressa, C. vernalis, C. vulpina, C. solstitialis, N. podagrica*

Angelica sylvestris, **wild angelica (35):** *M. mellinum, M. scalare* (Uffen, 1969), *P. albimanus* (Uffen, 1969), *P. scutatus* (Uffen, 1969), *P. rosarum, C. bicinctum, C. festivum, E. balteatus, E. luniger, L. glaucia, M. labiatarum, M. cinctella, S. scripta* (Uffen, 1969), *S. ribesii, S. torvus, S. vitripennis* (Uffen, 1969), *C. pagana, C. proxima, C. scutellata* (Uffen, 1969), *C. soror, R. campestris* (Uffen, 1969), *C. solstitialis, R. splendens, S. verecunda, E. arbustorum, E. horticola, E. interruptus, E. pertinax, E. tenax, M. florea, P. viduata, S. silentis* (Uffen, 1969), *V. inanis, V. zonaria, S. pipiens*

Pastinaca sativa, **wild parsnip (45):** *M. scalare, P. albimanus, P. clypeatus* s.s., *P. granditarsus, P. rosarum, C. bicinctum, C. elegans, C. verralli, E. grossulariae, E. balteatus, E. corollae, E. luniger* (LP, 1950a), *L. glaucia, M. compositarum, M. auricollis, M. cinctella, S. pyrastri, S. scripta, S. ribesii, S. vitripennis, X. pedissequum, C. barbata, C. illustrata, C. impressa, C. pagana, C. proxima, C. scutellata, C. soror, C. vernalis, C. vulpina, C. cemiteriorum, N. podagrica, O. nobilis, E. sepulchralis* (LP, 1950a), *E. arbustorum, E. interruptus, E. pertinax, E. tenax, H. pendulus, M. cimbiciformis* (LP, 1950a), *M. florea, P. noctiluca, P. viduata, T. primus, S. pipiens*

Heracleum sphondylium, **hogweed (67):** *M. mellinum, M. scalare,*
P. albimanus, P. scutatus, P. granditarsus, P. rosarum, C. bicinctum,
C. cautum, C. verralli, D. albostriatus, E. diaphana, E. grossulariae,
E. balteatus, E. corollae, E. latifasciatus, E. luniger, L. glaucia,
L. laternaria, L. lucorum, M. compositarum, M. labiatarum,
M. umbellatarum, M. guttatum, M. auricollis, M. cinctella, S. pyrastri,
S. taeniata, S. scripta, S. ribesii, S. torvus, S. vitripennis, X. pedissequum,
C. barbata, C. illustrata, C. impressa, C. pagana, C. proxima,
C. scutellata, C. soror, C. variabilis, C. vulpina, C. cemiteriorum,
C. solstitialis, M. dubia, N. podagrica, N. tenur, O, nobilis, R. splendens,
E. sepulchralis (LP, 1950a), *E. arbustorum, E. interruptus, E. pertinax,*
E. tenax, H. pendulus, M. florea, M. equestris, P. versicolor, P. austriaca,
P. noctiluca, P. viduata, P. virens, V. bombylans, V. inanis, V. inflata,
V. pellucens, V. zonaria, S. pipiens

Torilis japonica, **upright hedge-parsley (32):** *M. scalare, P. albimanus,*
P. angustatus, P. scutatus, C. bicinctum, C. festivum, E. diaphana,
E. balteatus, M. labiatarum, M. cinctella, S. pyrastri, S. scripta, S. ribesii,
S. vitripennis, C. bergenstammi, C. pagana, C. proxima, C. scutellata,
C. soror, C. cemiteriorum, C. solstitialis, M. dubia, N. podagrica,
O. nobilis, E. arbustorum, E. interruptus, E. pertinax, M. florea,
P. viduata, S. silentis, V. inanis, S. pipiens

Daucus carota, **wild carrot (29):** *P. rosarum, C. bicinctum, C. festivum,*
E. balteatus, E. corollae, E. luniger, M. compositarum, M. labiatarum,
S. scripta, S. ribesii, S. vitripennis, C. impressa, C. pagana, C. scutellata,
C. soror, C. vulpina, C. solstitialis, O. nobilis, R. splendens,
E. arbustorum, E. interruptus, E. pertinax, E. tenax, H. trivittatus,
M. florea, E. tuberculatus, P. viduata, P. virens, S. pipiens

Gentianaceae

Centaurium erythraea, **common centaury (1):** *S. scripta*

Convolvulaceae

Convolvulus arvensis, **field bindweed (23):** *M. mellinum, M. scalare,*
P. albimanus, C. bicinctum, C. festivum, E. grossulariae, E. balteatus,
E. corollae, E. luniger, S. pyrastri, S. scripta, S. ribesii, S. vitripennis,
X. pedissequum, F. cuprea, R. campestris, E. tenax, H. pendulus,
M. florea, P. versicolor, E. tuberculatus, M. equestris, S. pipiens

Calystegia sepium, **hedge bindweed (4):** *M. scalare, E. balteatus,*
F. cuprea, R. campestris

Calystegia silvatica, **large bindweed (2):** *E. balteatus, R. campestris*

Cuscutaceae

Cuscuta epithymum, **dodder (1):** *S. scripta*

Boraginaceae

Pentaglottis sempervirens, **green alkanet (1):** *R. campestris*

Myosotis scorpioides, **water forget-me-not (2):** *P. albimanus* (LP, 1950a), *P. rosarum* (LP, 1950a)

Myosotis **spp., forget-me-not (2):** *C. pagana, E. tenax*

Lamiaceae

Stachys sylvatica, **hedge woundwort (1):** *R. campestris*

Lamium album, **white dead-nettle (3):** *P. albimanus, L. lucorum, R. campestris*

Galeopsis tetrahit, **common hemp-nettle (1):** *R. campestris*

Ajuga reptans, **bugle (2):** *P. albimanus* (LP, 1950a), *R. campestris*

Glechoma hederacea, **ground-ivy (4):** *P. albimanus, R. campestris, E. arbustorum, H. pendulus* (LP, 1950a)

Prunella vulgaris, **self-heal (1):** *R. campestris* (LP, 1950a)

Clinopodium vulgare, **wild basil (1):** *P. albimanus, C. nemorum*

Origanum vulgare, **wild marjoram (18):** *C. festivum, C. verralli, E. balteatus, S. pyrastri, S. scripta, R. campestris, E. arbustorum, E. interruptus, E. intricarius, E. pertinax, E. tenax, H. hybridus, H. pendulus, H. trivittatus, M. florea, V. inanis, V. pellucens, S. pipiens*

Lycopus europaeus, **gipsywort (7):** *P. angustatus* (LP, 1950a), *E. luniger, S. scripta, E. sepulchralis* (LP, 1950a), *E. arbustorum* (LP, 1950a), *M. florea, S. pipiens*

Mentha arvensis, **corn mint (1):** *E. interruptus*

Mentha aquatica, **water mint (24):** *M. scalare* (LP, 1950a), *P. albimanus, P. peltatus, P. scutatus, C. bicinctum, C. festivum, E. balteatus, S. taeniata, S. vitripennis* (LP, 1950a), *R. campestris* (LP, 1950a), *E. sepulchralis, E. arbustorum, E. horticola, E. intricarius, E. interruptus* (LP, 1950a), *E. pertinax, E. tenax, H. hybridus, H. pendulus, M. florea, V. bombylans, V. inanis, V. pellucens, S. pipiens*

Lavandula **spp., garden lavender (1):** *R. campestris*

Plantaginaceae

Plantago lanceolata, **ribwort plantain (4):** *M. mellinum, M. scalare, C. festivum, E. luniger*

Buddlejaceae

Buddleja davidii, **butterfly-bush (11):** *E. balteatus, E. luniger, S. scripta, R. campestris, R. rostrata, E. interruptus, E. tenax, V. inanis, V. pellucens, V. zonaria, S. pipiens*

Oleaceae

Ligustrum ovalifolium, **garden privet (3):** *E. balteatus, E. pertinax, V. zonaria*

Ligustrum **spp., privet (2):** *E. horticola, E. interruptus*

Scrophulariaceae

Scrophularia nodosa, **common figwort (1):** *C. variabilis*

Scrophularia auriculata, **water figwort (1):** *C. variabilis*

Veronica chamaedrys **germander speedwell (3):** *P. haemorrhous, D. tricinctus* (Jennings, 1895), *C. pagana*

Veronica officinalis, **heath speedwell (1):** *S. scripta* (LP, 1950a)

Hebe **spp., hedge veronica (9):** *P. albimanus, E. balteatus, E. corollae, E. luniger, M. auricollis, E. pertinax, E. tenax, V. zonaria, S. pipiens*

Euphrasia **spp., eyebright (8):** *C. festivum, S. vitripennis, E. sepulchralis, E. arbustorum, E. intricarius, E. tenax, M. florea, S. pipiens*

Odontites vernus, **red bartsia (4):** *P. albimanus, E. balteatus, R. campestris, S. pipiens*

Campanulaceae

Campanula trachelium, **nettle-leaved bellflower (1):** *R. rostrata* (LP, 1956a)

Phyteuma orbiculare, **round-headed rampion (1):** *E. luniger*

Rubiaceae

Galium odoratum, **woodruff (1):** *C. albitarsis*

Galium palustre, **common marsh bedstraw (3):** *C. bicinctum, S. vitripennis* (LP, 1950a), *P. viduata*

Galium verum, **lady's bedstraw (3):** *E. balteatus, S. interrupta, S. scripta*

Galium saxatile, **heath bedstraw (1):** *S. scripta*

Galium aparine, **cleavers (1):** *S. pipiens*

Caprifoliaceae

Sambucus nigra, **elder (6):** *D. fasciata, E. grossulariae, E. balteatus, M. cinctella* (LP, 1950a), *S. ribesii, V. inflata* (LP, 1950a)

Viburnum lantana, **wayfaring tree (1):** *P. tarsalis* (LP, 1941)

Lonicera periclymenum **honeysuckle (1):** *M. scalare*

Valerianaceae

Valeriana officinalis, **common valerian (1):** *E. arbustorum*

Dipsacaceae

Knautia arvensis, field scabious (1): *E. grossulariae*

Succisa pratensis, devil's-bit scabious (20): *M. scalare, E. grossulariae, E. balteatus, S. vitripennis* (LP, 1941), *R. campestris, R. rostrata* (LP, 1950a), *E. arbustorum* (LP, 1950a), *E. horticola* (LP, 1941), *E. interruptus, E. intricarius, E. pertinax* (LP, 1950a), *E. tenax, M. florea, H. pendulus, H. trivittatus, S. silentis, V. inanis, V. pellucens* (LP, 1941), *V. zonaria, S. pipiens* (LP, 1950a)

Scabiosa columbaria, small scabious (5): *E. grossulariae, E. balteatus, E. intricarius, E. pertinax, E. tenax*

Asteraceae

Arctium lappa, greater burdock (1): *S. pipiens*

Carduus crispus, welted thistle (3): *P. albimanus, E. balteatus, R. campestris*

Cirsium vulgare, spear thistle (4): *D. albostriatus, E. balteatus, S. pyrastri, R. campestris*

Cirsium palustre, marsh thistle (4): *E. balteatus, E. intricarius* (LP, 1950a), *E. pertinax* (LP, 1950a), *E. tenax*

Cirsium arvense, creeping thistle (20): *M. scalare, C. bicinctum, E. balteatus, E. corollae, S. scripta, S. ribesii, F. cuprea, E. arbustorum, E. interruptus, E. intricarius, E. pertinax, E. tenax, H. hybridus, H. pendulus, H. trivittatus, M. florea, V. inanis, V.pellucens, V. zonaria, S. pipiens*

Cirsium/Carduus spp., thistle spp.(1): *M. dubia* (Withers, 1983)

Centaurea scabiosa, greater knapweed (1): *E. grossulariae*

Centaurea nigra, common knapweed (18): *C. bicinctum, C. festivum, D. albostriatus, E. grossulariae, E. balteatus, E. luniger, S. pyrastri, S. ribesii, S. vitripennis, R. campestris, E. arbustorum* (LP, 1956b), *E. interruptus, E. intricarius, E. tenax, H. pendulus, H. trivittatus* (LP, 1956b), *V. pellucens, S. pipiens*

Lapsana communis, nipplewort (7): *M. mellinum* (LP, 1950a), *P. albimanus, E. balteatus, M. auricollis, S. ribesii, F. cuprea, S. pipiens*

Hypochaeris radicata, cat's-ear (18): *M. scalare, P. albimanus, P. scutatus, D. albostriatus, E. balteatus, E. corollae* (LP, 1956b), *E. latifasciatus, E. luniger, M. auricollis, S. scripta, S. ribesii* (LP, 1956b), *S. vitripennis, N. podagrica, E. arbustorum* (LP, 1956b), *E. tenax, P. tricincta, S. silentis, S. pipiens*

Leontodon autumnalis, autumn hawkbit (11): *P. albimanus, C. bicinctum, D. albostriatus, E. balteatus, E. luniger, M. auricollis, S. scripta, S. taeniata, S. ribesii, S. vitripennis, E. tenax*

Leontodon hispidus, **rough hawkbit (5):** *P. albimanus, S. scripta, E. arbustorum, E. tenax, M. florea*

Picris echioides, **bristly oxtongue (4):** *E. balteatus, S. vitripennis, E. tenax, H. pendulus*

Picris hieracioides, **hawkweed oxtongue (7):** *C. bicinctum, D. albostriatus, E. balteatus, E. luniger, S. scripta, F. cuprea, H. pendulus*

Sonchus arvensis, **perennial sow-thistle (15):** *M. mellinum, P. albimanus, P. scutatus, D. albostriatus, E. balteatus, S. pyrastri* (LP, 1950a), *S. ribesii, C. pagana, F. cuprea, E. arbustorum, E. pertinax, E. tenax, H. pendulus, M. florea, S. pipiens*

Sonchus oleraceus, **smooth sow-thistle (2):** *P. scutatus, E. balteatus*

Sonchus asper, **prickly sow-thistle (6):** *M. mellinum* (LP, 1950a), *P. albimanus* (LP, 1950a), *E. balteatus* (LP, 1950a), *E. sepulchralis* (LP, 1950a), *E. tenax* (LP, 1950a), *H. pendulus* (LP, 1950a)

Taraxacum **spp., dandelion (24):** *M. scalare, P. albimanus, P. angustatus, P. scutatus, P. tarsalis, D. albostriatus, D. tricinctus, D. venustus, E. balteatus, L. lucorum, M. cincta, S. scripta, S. ribesii, S. torvus* (LP, 1950a), *S. vitripennis, X. pedissequum, C. bergenstammi, C. pagana, F. cuprea, R. campestris, E. arbustorum, E. pertinax, E. tenax, H. pendulus*

Pilosella officinarum, **mouse-ear hawkweed (2):** *P. haemorrhous, C. bicinctum*

Hieracium **spp., hawkweed spp. (5):** *S. scripta, E. sepulchralis, E. arbustorum, E. tenax, M. florea*

Crepis capillaris, **smooth hawk's-beard (11):** *C. festivum, E. balteatus, E. luniger, S. scripta, S. ribesii, S. vitripennis, F. cuprea, E. arbustorum, E. tuberculatus, M. equestris, S. pipiens*

Crepis vesicaria, **beaked hawk's-beard (3):** *E. corollae, M. equestris, P. viduata*

Pulicaria dysenterica, **common fleabane (25):** *M. scalare, P. albimanus, P. scutatus, P. granditarsus, C. bicinctum, E. balteatus, E. corollae, E. latifasciatus, S. pyrastri, S. scripta, S. taeniata, S. ribesii, S. vitripennis, C. bergenstammi, C. fraterna, C. proxima, C. vernalis, E. sepulchralis, E. arbustorum, E. interruptus, E. pertinax, E. tenax, H. pendulus, M. florea, S. pipiens*

Solidago virgaurea, **goldenrod (1):** *M. scalare*

Solidago canadensis, **Canadian goldenrod (11):** *M. scalare, E. latifasciatus, C. scutellata, E. arbustorum, E. interruptus, E. pertinax, E. tenax, H. pendulus, M. florea, V. inanis, S. pipiens*

Aster spp., **Michaelmas-daisy (10):** *C. festivum, E. balteatus, S. scripta, E. sepulchralis, E. arbustorum, E. intricarius, E. tenax, H. pendulus, M. florea, S. pipiens*

Tanacetum vulgare, **tansy (1):** *E. balteatus*

Achillea ptarmica, **sneezewort (6):** *P. angustatus* (LP, 1950a), *S. scripta, E. arbustorum* (LP, 1950a), *E. tenax, H. hybridus* (LP, 1950a), *S. pipiens*

Achillea millefolium, **yarrow (18):** *C. bicinctum, E. balteatus, M. cinctella, S. scripta, C. latifrons, C. vernalis, C. solstitialis, O. nobilis, E. sepulchralis* (LP, 1956b), *E. arbustorum, E. interruptus, E. pertinax, E. tenax, H. pendulus, H. trivittatus* (LP, 1956b), *M. florea* (LP, 1956b), *V. inanis, S. pipiens*

Leucanthemum vulgare, **oxeye daisy (12):** *C. bicinctum, E. luniger, S. ribesii, E. arbustorum, E. horticola, E. interruptus, E. pertinax, E. tenax, H. pendulus, H. trivittatus, M. equestris, S. pipiens*

Tripleurospermum inodorum, **scentless mayweed (19):** *P. albimanus, P. scutatus, P. granditarsus, E. balteatus, E. corollae, S. scripta, S. taeniata, S. vitripennis, L. metallina, N. podagrica, E. sepulchralis, E. arbustorum, E. interruptus, E. intricarius, E. pertinax, E. tenax, H. pendulus, M. florea, S. pipiens*

Senecio jacobaea, **common ragwort (21):** *M. scalare, P. scutatus, E. balteatus, M. auricollis, S. pyrastri, S. scripta, S. ribesii, S. vitripennis, C. bergenstammi, C. fraterna, C. scutellata, E. sepulchralis, E. arbustorum, E. interruptus, E. pertinax, E. tenax, H. pendulus, P. noctiluca, S. silentis, V. inanis, S. pipiens*

Senecio erucifolius, **hoary ragwort (15):** *C. festivum, D. albostriatus, E. balteatus, E. luniger* (LP, 1950a), *S. scripta, S. vitripennis, E. sepulchralis, E. arbustorum, E. intricarius, E. pertinax, E. tenax, H. pendulus, H. trivittatus, M. florea, S. pipiens*

Senecio squalidus, **Oxford ragwort (1):** *H. pendulus*

Senecio spp., **ragwort spp. (25):** *M. scalare, P. albimanus, E. balteatus, M. cinctella, S. pyrastri, S. scripta, S. taeniata, S. ribesii, S. vitripennis, C. bergenstammi, C. pagana, R. campestris, C. solstitialis, E. sepulchralis, E. arbustorum, E. horticola, E. interruptus, E. pertinax, E. tenax, H. pendulus, H. trivittatus, M. florea, S. silentis, V. pellucens, S. pipiens*

Tussilago farfara, **colt's-foot (2):** *C. grossa, C. pagana*

Eupatorium cannabinum, **hemp-agrimony (3):** *E. interruptus, E. pertinax, S. pipiens*

Alismataceae

Alisma plantago-aquatica, **water-plantain (12):** *P. albimanus, P. scutatus, P. granditarsus, E. balteatus, S. taeniata, C. pagana, C. proxima, A. lineata, E. sepulchralis, E. pertinax, M. florea, S. pipiens*

Liliaceae

Hyacinthoides non-scripta, **bluebell (9):** *M. mellinum* (LP, 1950a), *M. scalare* (LP, 1955b), *P. albimanus* (LP, 1950a), *P. scutatus* (LP, 1955b), *P. tarsalis* (LP, 1955b), *D. venustus* (LP, 1955b), *E. luniger* (LP, 1955b), *M. lasiophthalma, R. campestris* (LP, 1950a)

Allium ursinum, **ramsons (5):** *P. albimanus, P. scutatus, P. maculata, E. pertinax, C. asilica*

Iridaceae

Iris pseudacorus, **yellow iris (1):** *P. versicolor*

Dioscoreaceae

Tamus communis, **black bryony (1):** *Baccha* spp.

APPENDIX 3 – Index of plants listed by common name

agrimony, *Agrimonia eupatoria*
alder buckthorn, *Frangula alnus*
apple, *Malus* spp.
ash, *Fraxinus excelsior*
autumn hawkbit, *Leontodon autumnalis*

bastard cabbage, *Rapistrum rugosum*
beaked hawk's-beard, *Crepis vesicaria*
beech, *Fagus sylvatica*
bell heather, *Erica cinerea*
bilberry, *Vaccinium myrtillus*
birch, *Betula pendula/pubescens*
bird cherry, *Prunus padus*
black bryony, *Tamus communis*
black medick, *Medicago lupulina*
blackthorn, *Prunus spinosa*
bluebell, *Hyacinthoides non-scripta*
bracken, *Pteridium aquilinum*
bramble, *Rubus fruticosus* agg.
branched bur-reed, *Sparganium erectum*
bristly oxtongue, *Picris echioides*
broad-leaved dock, *Rumex obtusifolius*
broom, *Cytisus scoparius*
buckwheat, *Fagopyrum esculentum*
bugle, *Ajuga reptans*
bulbous buttercup, *Ranunculus bulbosus*
bulrush, *Typha latifolia*
burnet-saxifrage, *Pimpinella saxifraga*
butterbur, *Petasites hybridus*
buttercup spp., *Ranunculus* spp.
butterfly-bush, *Buddleja davidii*

Canadian goldenrod, *Solidago canadensis*
cat's-ear, *Hypochaeris radicata*
charlock, *Sinapis arvensis*
cherry laurel, *Prunus laurocerasus*
chickweed, *Stellaria media*
cleavers, *Galium aparine*
colt's-foot, *Tussilago farfara*
common centaury, *Centaurium erythraea*
common figwort, *Scrophularia nodosa*
common fleabane, *Pulicaria dysenterica*
common hemp-nettle, *Galeopsis tetrahit*
common knapweed, *Centaurea nigra*

common marsh bedstraw, *Galium palustre*
common nettle, *Urtica dioica*
common poppy, *Papaver rhoeas*
common ragwort, *Senecio jacobaea*
common reed, *Phragmites australis*
common rock-rose, *Helianthemum nummularium*
common valerian, *Valeriana officinalis*
corn mint, *Mentha arvensis*
cow parsley, *Anthriscus sylvestris*
creeping buttercup, *Ranunculus repens*
creeping cinquefoil, *Potentilla reptans*
creeping thistle, *Cirsium arvense*
creeping willow, *Salix repens*
cuckoo-flower, *Cardamine pratensis*

dandelion, *Taraxacum* spp.
devil's-bit scabious, *Succisa pratensis*
dodder, *Cuscuta epithymum*
dog rose, *Rosa canina*
dog's mercury, *Mercuralis perennis*
dogwood, *Cornus sanguinea*
dropwort, *Filipendula vulgaris*

elder, *Sambucus nigra*
elm, *Ulmus* spp.
enchanter's nightshade, *Circaea lutetiana*
English elm, *Ulmus procera*
eyebright, *Euphrasia* spp.

field maple, *Acer campestre*
field scabious, *Knautia arvensis*
figwort, *Scrophularia* spp.
fir, *Abies* spp.
firethorn, *Pyracantha coccinea*
fool's water-cress, *Apium nodiflorum*
forget-me-not, *Myosotis* spp.

garden lavender, *Lavandula* spp.
garden privet, *Ligustrum ovalifolium*
garlic mustard, *Alliaria petiolata*
germander speedwell, *Veronica chamaedrys*
gipsywort, *Lycopus europaeus*
goat willow, *Salix caprea*
goldenrod, *Solidago virgaurea*
goosefoot, *Chenopodium* spp.
great horsetail, *Equisetum telmateia*
great willowherb, *Epilobium hirsutum*
greater burdock, *Arctium lappa*

greater knapweed, *Centaurea scabiosa*
greater stitchwort, *Stellaria holostea*
green alkanet, *Pentaglottis sempervirens*
grey poplar, *Populus canescens*
ground-elder, *Aegopodium podagraria*
ground-ivy, *Glechoma hederacea*

hawkweed oxtongue, *Picris hieracioides*
hawkweed spp., *Hieracium* spp.
hawthorn, *Crataegus* spp.
heath speedwell, *Veronica officinalis*
heather, *Calluna vulgaris*
hedge bindweed, *Calystegia sepium*
hedge mustard, *Sisymbrium officinale*
hedge veronica, *Hebe* spp.
hedge woundwort, *Stachys sylvatica*
hemlock, *Conium maculatum*
hemlock water-dropwort, *Oenanthe crocata*
hemp-agrimony, *Eupatorium cannabinum*
herb robert, *Geranium robertianum*
hoary ragwort, *Senecio erucifolius*
hogweed, *Heracleum sphondylium*
holly, *Ilex aquifolium*
honeysuckle, *Lonicera periclymenum*
horse chestnut, *Aesculus hippocastanum*
horse-radish, *Armoracia rusticana*
hybrid lime, *Tilia x europaea*

imperforate St John's-wort, *Hypericum maculatum*
Indian balsam, *Impatiens glandulifera*
ivy, *Hedera helix*

Japanese knotweed, *Fallopia japonica*

kidney vetch, *Anthyllis vulneraria*
knotgrass, *Polygonum aviculare*

lady's bedstraw, *Galium verum*
large bindweed, *Calystegia silvatica*
lesser celandine, *Ranunculus ficaria*
lesser spearwort, *Ranunculus flammula*
lesser stitchwort, *Stellaria graminea*

marsh marigold, *Caltha palustris*
marsh thistle, *Cirsium palustre*
meadow buttercup, *Ranunculus acris*
meadowsweet, *Filipendula ulmaria*
melilot, *Melilotus* spp.
Michaelmas-daisy, *Aster* spp.

mouse-ear hawkweed, *Pilosella officinarum*
mugwort, *Artemisia vulgaris*
musk thistle, *Carduus nutans*

navelwort, *Umbilicus rupestris*
nettle-leaved bellflower, *Campanula trachelium*
nipplewort, *Lapsana communis*
Norway maple, *Acer platanoides*

oak, *Quercus* spp.
opium poppy, *Papaver somniferum*
orpine, *Sedum telephium*
oxeye daisy, *Leucanthemum vulgare*
Oxford ragwort, *Senecio squalidus*

pepper-saxifrage, *Silaum silaus*
perennial sow-thistle, *Sonchus arvensis*
perennial wall-rocket, *Diplotaxis tenuifolia*
perforate St John's-wort, *Hypericum perforatum*
pignut, *Conopodium majus*
pine, *Pinus* spp.
plantain, *Plantago* spp.
prickly sow-thistle, *Sonchus asper*
primrose, *Primula vulgaris*
privet, *Ligustrum* spp.

ragged robin, *Lychnis flos-cuculi*
ragwort spp., *Senecio* spp.
ramsons, *Allium ursinum*
rape, *Brassica napus*
raspberry, *Rubus idaeus*
red bartsia, *Odontites vernus*
red clover, *Trifolium pratense*
redshank, *Persicaria maculosa*
reed sweet-grass, *Glyceria maxima*
rhododendron, *Rhododendron ponticum*
ribwort plantain, *Plantago lanceolata*
rosebay willowherb, *Chamerion angustifolium*
round-headed rampion, *Phyteuma orbiculare*
rough chervil, *Chaerophyllum temulum*
rough hawkbit, *Leontodon hispidus*
rowan, *Sorbus aucuparia*
rush, *Juncus* spp.

sallow, *Salix caprea/cinerea*
sand spurrey, *Spergularia rubra*
sanicle, *Sanicula europaea*
scentless mayweed, *Tripleurospermum inodorum*

Scots pine, *Pinus sylvestris*
self-heal, *Prunella vulgaris*
silverweed, *Potentilla anserina*
small scabious, *Scabiosa columbaria*
smooth hawk's-beard, *Crepis capillaris*
smooth sow-thistle, *Sonchus oleraceus*
sneezewort, *Achillea ptarmica*
spear thistle, *Cirsium vulgare*
square-stalked willowherb, *Epilobium tetragonum*
St Dabeoc's heath, *Daboecia cantabrica*
sticky mouse-ear, *Cerastium glomeratum*
stinking goosefoot, *Chenopodium vulvaria*
sweet chestnut, *Castanea sativa*
sycamore, *Acer pseudoplatanus*

tansy, *Tanacetum vulgare*
thistle spp., *Cirsium/Carduus* spp.
tormentil, *Potentilla erecta*
traveller's joy, *Clematis vitalba*

upright hedge-parsley, *Torilis japonica*

violet, *Viola* spp.

water chickweed, *Myosoton aquaticum*
water-cress, *Rorippa nasturtium-aquaticum*
water crowfoot, *Ranunculus*, subgen. *Batrachium*
water figwort, *Scrophularia auriculata*
water forget-me-not, *Myosotis scorpioides*
water mint, *Mentha aquatica*
water-plantain, *Alisma plantago-aquatica*
weld, *Reseda luteola*
welted thistle, *Carduus crispus*
white bryony, *Bryonia dioica*
white dead-nettle, *Lamium album*
white mignonette, *Reseda alba*
white poplar, *Populus alba*
wild angelica, *Angelica sylvestris*
wild basil, *Clinopodium vulgare*
wild cabbage, *Brassica oleracea*
wild carrot, *Daucus carota*
wild cherry, *Prunus avium*
wild marjoram, *Origanum vulgare*
wild parsnip, *Pastinaca sativa*
willow, *Salix* spp. other than *S. caprea/cinerea*
winter-cress, *Barbarea vulgaris*
wood anemone, *Anemone nemorosa*
wood avens, *Geum urbanum*

wood dock, *Rumex sanguineus*
wood spurge, *Euphorbia amygdaloides*
woodruff, *Galium odoratum*

yarrow, *Achillea millefolium*
yellow iris, *Iris pseudacorus*
yellow loosestrife, *Lysimachia vulgaris*
yellow pimpernel, *Lysimachia nemorum*

APPENDIX 4 – Species list for
Great Bookham Common

Baccha spp

Melanostoma mellinum

M. scalare

Platycheirus albimanus

P. ambiguus

P. discimanus

P. manicatus

P. scutatus

P. tarsalis

P. granditarsus

P. rosarum

Chrysotoxum bicinctum

C. cautum

C. elegans

C. festivum

C. verralli

Dasysyrphus albostriatus

D. tricinctus

D. venustus

Epistrophe diaphana

E. eligans

E. grossulariae

E. melanostoma

E. nitidicollis

Episyrphus balteatus

Eupeodes corollae

E. latifasciatus

E. luniger

E. nitens

Leucozona glaucia

L. laternaria

L. lucorum

Melangyna barbifrons

M. cincta

M. compositarum

M. labiatarum

M. lasiophthalma

M. umbellatarum

Meligramma euchromum

M. trianguliferum

Meliscaeva auricollis

M. cinctella

Parasyrphus lineola

P. punctulatus

Scaeva pyrastri

S. selenitica

Sphaerophoria rueppellii

S. scripta

S. taeniata

Syrphus ribesii

S. torvus

S. vitripennis

Xanthogramma citrofasciatum

X. pedissequum

Cheilosia albipila

C. albitarsis

C. barbata

C. carbonaria

C. fraterna

C. grossa

C. illustrata

C. impressa

C. lasiopa

C. latifrons

C. nebulosa

C. pagana

C. praecox

C. proxima

C. scutellata

C. soror

C. variabilis

C. vernalis

C. vulpina

Ferdinandea cuprea

Rhingia campestris

R. rostrata

Brachyopa insensilis

Chrysogaster cemiteriorum

C. solstitialis

Melanogaster hirtella

Myolepta dubia

Neoascia podagrica
N. geniculata
N. interrupta
N. meticulosa

Orthonevra nobilis

Riponnensia splendens

Anasimyia contracta
A. lineata
A. transfuga

Eristalinus sepulchralis

Eristalis arbustorum
E. horticola
E. interruptus
E. intricarius
E. pertinax
E. tenax

Helophilus hybridus
H. pendulus
H. trivittatus

Mallota cimbiciformis

Myathropa florea

Parhelophilus frutetorum
P. versicolor

Eumerus ornatus
E. tuberculatus

Merodon equestris

Heringia pubescens
H. vitripennis

Pipiza austriaca
P. bimaculata
P. fenestrata
P. luteitarsis
P. noctiluca

Pipizella viduata
P. virens

Sericomyia silentis

Volucella bombylans
V. inanis
V. inflata
V. pellucens

Brachypalpoides lentus

Brachypalpus laphriformis

Chalcosyrphus nemorum

Criorhina asilica
C. berberina
C. floccosa
C. ranunculi

Syritta pipiens

Xylota abiens
X. segnis
X. sylvarum

Microdon mutabilis

Probable misidentifications

Platycheirus scambus
(Parmenter, 1966)

Cheilosia vicina
(Parmenter, 1950a)

APPENDIX 5 – Organisations

Study societies

British Entomological and Natural History Society
(Publisher of *British Hoverflies*, by A. Stubbs)
Dinton Pastures Country Park, Davis Street, Hurst,
Reading, Berks RG10 0TH.

Dipterists' Forum
(Publisher of the Hoverfly Newsletter)
Membership: Dr E. Howe, Ger-y-parc, Brynteg, Benllech,
Anglesey, WL74 8NS

Amateur Entomologists' Society
P.O. Box 8774, London SW7 5ZG.

Croydon Natural History and Scientific Society
96a Brighton Road, South Croydon, Surrey CR2 6AD.

Conservation bodies

Surrey Wildlife Trust
School Lane, Pirbright, Woking, Surrey GU24 0JN.

APPENDIX 6 – Bibliography

These lists include those texts directly referred to in this book and also those used as literature sources for the maps. The list is not exhaustive as I have not been able to dedicate sufficient time to a full literature search, but it does offer an indication of the range of sources available.

Adams, F.C., 1899
>*Brachypalpus laphriformis* Mcq., in the New Forest & c. *Entomologist's mon. Mag.* **35**: 71

Airy Shaw, H.K., 1955
>*Volucella zonaria* Poda and *Vespa crabro* L. at Kew. *Amateur Entomologists' Society Bulletin* **14**: 39

Baker, P.J., 1974
>BENHS field meeting: Chobham Common, Surrey 15th July 1973. *Proc. Brit. ent. nat. Hist. Soc.* **6** (4): 121

Baker, P.J., 1984
>BENHS field meeting: Chobham Common, Surrey 2.vii.1983. *Proc. Brit. ent. nat. Hist. Soc.* **17** (1/2): 61-62

Baker, P.J., 1986
>BENHS field meeting: Chobham Common, Surrey 24.v.1986. *Proc. Brit. ent. nat. Hist. Soc.* **19** (1/4): 66-67

Balfour-Browne, J., 1949
>*Volucella zonaria* Poda (Dipt., Syrphidae) in Surrey. *Entomologist's mon. Mag.* **82**: 303

Ball, S.G., & Morris, R.K.A., 1992
>Progress report 1, March 1992. Hoverfly Newsletter No. 14.

Ball, S.G., & Morris R.K.A. (in press)
>*A provisional atlas of British hoverflies (Diptera, Syrphidae).* Institute of Terrestrial Ecology, Monks Wood.

Beuk, P.L.T., 1989
>BENHS indoor meeting 27.vii.1989. *Br. J. Ent. nat. Hist.* **2** (4): 192

Beuk, P.L.T., 1990
>A hoverfly of the genus *Epistrophe* (Dipt., Syrphidae) new to Britain. *Entomologist's mon. Mag.* **126**: 167-170

Billups, T.R., 1891
>Two and a half hours' investigation of the entomology of Oxshott. *Entomologist* **24**: 201-204

Blair, K.G., 1945
>Pairing of *Penthesilea ranunculi* Panz. (Dipt. Syrphidae). *The Entomologist* **78**: 127

Burton, R.M., 1983
>*Flora of the London Area.* London Natural History Society.

Buck de, N., 1990
>Bloembezoek en bestuivingsecologie van Zweefvliegen (Diptera, Syrphidae) in het bijzonder voor België. *Documents de travail de l'Institut royal des Sciences naturelles de Belgique 60.* Brussels

Champion, G.C., 1912
>*Syrphus torvus* O-S, and *S. luniger*, Meig., bred. *Entomologist's mon. Mag.* **48**: 215-216

Chandler, P.J., 1969
>The hover-flies of Kent. *Trans. Kent field club* **3** (3): 139-202

Chandler, P.J., (in press)
>A Checklist of the Insects of the British Isles (New Series) part 1: Diptera. *Handbk. Ident. Br. Insects* **XII** (1). Royal Entomological Society of London.

Claussen, C., 1980
>Die Schwebfliegenfauna des Landesteils Schleswig in Schleswig-Holstein (Diptera, Syrphidae). *Supplement zu Faunistisch-Ökologische Mitteilungen* **1**: 3-79

Coe, R.L., 1939a
>Description of the female of *Xylota xanthocnema* Collin (Dipt., Syrphidae). *Entomologist's mon. Mag.* **75**: 224

Coe, R.L., 1939b
>A second British record of *Rhingia rostrata* Linnaeus (Dipt. Syrphidae); its distinctions from *R. campestris* Meigen. *Entomologist's mon. Mag.* **75**: 225-227

Coe, R.L., 1954
>Diptera at Oxshott, Surrey. Entomologist **87**:116

Coe, R.L., 1961
>Massed occurrence of the rare syrphid fly, *Rhingia rostrata* Linnaeus at Selsdon Woods in Surrey. *Entomologist* **94**: 257

Collin, J.E., 1918
>A Dipteron new to the British List. *Trans. Ent. Soc. Lond.* (Proc.) 1918: lxxvii

Collin, J.E., 1940
Notes on Syrphidae (Diptera) IV. *Entomologist's mon. Mag.*
76:150-157

Collins, G.A., 1997
Larger Moths of Surrey. Surrey Wildlife Trust

Cowley, J., 1949
Some Diptera from Surrey and the South-west of England.
J. Soc. Br. Ent. **3**: 101-118

Dandy, J.E., 1969
Watsonian vice-counties of Great Britain. Ray Society, London

Danks, H.V., 1963
A note on *Volucella zonaria. Amateur Entomologists' Society
Bulletin* **22**: 140

DoE, 1994
Biodiversity, the UK Action Plan. HMSO, London.

DoE, 1995
Biodiversity: the UK Steering Group Report. HMSO, London.

Dobson, J., 1992
Sphaerophoria rueppellii: further plant associations. Hoverfly
Newsletter No 15.

Dobson, J., 1997
Oviposition in *Epistrophe diaphana* (Syrphidae). *Dipterists
Digest* (new series) **4** (1): 47

Donisthorpe, H., 1927
The guests of British ants. London.

Drucker, G.R.F., Whitbread, A., & Barton, J., 1988
Surrey inventory of ancient woodlands (provisional). Nature
Conservancy Council, Peterborough. Unpublished.

Dunn, R.D., & Johnson, C.M., 1984
Diptera *in* The ecology of Mitcham Common edited by R.K.A.
Morris. Report for the Board of Conservators of Mitcham
Common.

Eagles, T.R., 1946
South London Entomological and Natural History Society:
August 28th, 1946. *Entomologist's mon. Mag.* **82**: 255

Eagles, T.R., 1948
South London Entomological and Natural History Society:
September 8th, 1948. *Entomologist's mon. Mag.* **84**: xl.

Eagles, T.R., 1949
South London Entomological and Natural History Society:
September 14th, 1949. *Entomologist's mon. Mag.* **85**: 264

Eagles, T.R., 1950
South London Entomological and Natural History Society:
June 14th, 1950. *The Entomologist* **83**: 191-192

Eagles, T.R., 1954
South London Entomological and Natural History Society:
April 28th, 1954. *Entomologist's mon. Mag.* **90**: xxiii

Eagles, T.R., 1955
SLENHS field meeting: Bookham Common 15th August 1953.
Proc. S. Lond. ent. nat. Hist. Soc. **1953-54**. 87-89

Edwards, F.W., 1923
Note on some British species of *Microdon* (Diptera,
Syrphidae). *Entomologist's mon. Mag.* **59**: 233-234

Evans, K.G.W., 1974
BENHS field meeting: Ashtead Common, Surrey 26th May
1973. *Proc. Brit. ent. nat. Hist. Soc.* **6** (4): 113-114

Falk, S.J., 1991
A review of the scarce and threatened flies of Great Britain
(Part 1). *Research & Survey in Nature Conservation* **39**.
Nature Conservancy Council, Peterborough.

Fincham Turner, J., 1949
Volucella zonaria, Poda. *Entomologist's Rec. J. Var.* **61**: 131

Follett, P., 1996
Dragonflies of Surrey. Surrey Wildlife Trust.

Foster, A.P., 1987
Xanthogramma pedissequum (Harris) (Dipt., Syrphidae) bred
from *Lasius niger* L. (Hym., Formicidae) nest. *Entomologists
Rec. J. Var.* **99**: 153-155

Fraser, F.C., 1955
Volucella zonaria Poda an established insect. *Amateur
Entomologists' Society Bulletin* **14**: 38-39

Fry, R., 1997
Finding *Criorhina ranunculi*. Hoverfly Newsletter No. 24

Fry, R., and Denton, J., 1992
A late date for *Chrysogaster virescens*, in Surrey. Hoverfly
Newsletter No. 15

Fry, R., and Lonsdale, D. (eds.), 1991
Habitat conservation for insects - a neglected green issue.
Amateur Entomologist **21**

Gilbert, F.S., 1986
Hoverflies. Naturalists Handbook No. 5. Cambridge University Press.

Goeldlin de Tiefenau, P., Maibach, A., & Speight, M.C.D., 1990
Sur quelques espèces de *Platycheirus* (Diptera, Syrphidae) nouvelles ou méconnues. *Dipterists Digest* **5**: 19-44

Goffe, E.R., 1945
Volucella zonaria (Poda, 1761) (Dipt. Syrphidae) in Britain. *Entomologist's mon. Mag.* **81**: 159-162

Greenwood, J.A.C., 1967(a)
SLENHS field meeting: Cosford Mill, Surrey 21st May 1967. *Proc. S. Lond. ent. nat. Hist. Soc.* **1967**. 89

Greenwood, J.A.C., 1967(b)
SLENHS field meeting:Thursley, Surrey 23rd July 1967. *Proc. S. Lond. ent. nat. Hist. Soc.* **1967**. 117-118

Greenwood, J.A.C., 1969
BENHS field meeting: Cosford Mill, Thursley, Surrey 19th May 1968. *Proc. Brit. ent. nat. Hist. Soc.* **2** (1): 20-21

Groves, E.W., 1956
On the egg laying of some Syrphidae. *Entomologists Rec. J. Var.* **68**: 275

Halstead, A.J., 1986
BENHS indoor meeting 12.vi.1986 - exhibit. *Proc. Brit. ent. nat. Hist. Soc.* **19**: 78

Hammond, C.O., 1971
BENHS indoor meeting - proceedings 8th July 1971. *Proc. Brit. ent. nat. Hist. Soc.* **4** (3): 94-95

Hastings, R., 1988
Winter-active hoverflies at Kew Gardens. Hoverfly Newsletter 8.

Hawkins, R.D., 1985
BENHS indoor meeting - 27.vi.1985. *Proc. Brit. ent. nat. Hist. Soc.* **18**: 81

Haynes, R.F., 1954
SLENHS field meeting: Bagshot, Surrey 23rd August 1953. *Proc. S. Lond. ent. nat. Hist. Soc.* **1953-54**. 89-90

Hobby, B.M., & Smith, K.G.V., 1961
The bionomics of *Empis tessellata* F. (Dipt., Empididae).
Entomologist's mon. Mag. **97**: 2-10

Holland, P.C., & Eagles, T.R., 1967
SLENHS field meeting: Oxshott, Surrey 15th October 1966.
Proc. S. Lond. ent. nat. Hist. Soc. **1967**. 26-28

Hughs, M.O., 1964
Letters to the Editor. *Amateur Entomologists' Society Bulletin*
23: 141

Iliff, D., 1991
Sphaerophoria rueppellii: plant associations. Hoverfly
Newsletter No. 13

Ingham, J.D., 1971
BENHS proceedings: 25th June 1970 - exhibit. *Proc. Brit. ent.*
nat. Hist. Soc. **4** (1): 1

Jennings, F.B., 1895
Didea fasciata Mcq. *Entomologist's mon. Mag.* **31**: 280-281

Jones, A.W., 1954
Notes on the drone-flies (Syrphidae, Diptera) of Wimbledon
Common. *London Naturalist* **33**: 83-88

Jones, A.W., 1964
Pocota personata Harris and *Criorhina* spp. (Diptera:
Syrphidae) in the London area. *Entomologist's Rec. J. Var.* **76**:
174-175

Jones, R.A., 1995
BENHS field meeting: Nunhead Cemetery, London SE15,
9.vii.1994. *Br. J. Ent. nat. Hist.* **8** (2): 92-93

Kirby, P., 1992
Habitat management for invertebrates: a practical handbook.
RSPB, Sandy.

Kirkpatrick, T.W., 1918
Diptera in 1917. *Entomologist's mon. Mag.* **54**: 18

Lees, D., 1990
A rearing record for *Didea fasciata*. Hoverfly Newsletter No. 11.

Lees, D., 1991
BENHS exhibition report - exhibit. *Br. J. Ent. nat. Hist.* **3** (2): 81

Leston, D., 1955
SLENHS field meeting: Box Hill 12th September 1953. *Proc. S. Lond. ent. nat. Hist. Soc.* **1953-54**. 91-92

Levy, D.A., Levy, E.T., & Dean, W.F., 1992
Dorset Hoverflies. Dorset Biological Records Centre

Lousley, J.E., 1976
Flora of Surrey. David and Charles, London.

McLean, I.F.G., & Stubbs, A.E., 1990
The breeding site of *Brachyopa pilosa* (Diptera; Syrphidae). *Dipterists Digest* **3**: 40

Messenger, J.L., 1970
BENHS field meeting: Cosford Mill, Surrey 11th May 1969. *Proc. Brit. ent. nat. Hist. Soc.* **3** (1): 26-27

Miles, S.R., 1989
BENHS annual exhibition 1988. BENHS Exhibition report. *Br. J. Ent. nat. Hist.* **2** (1): 46

Moore, B.P., 1957
SLENHS field meeting: Eashing Moors 12th August 1956. *Proc. S. Lond. ent. nat. Hist. Soc.* **1956**. 82-83

Moore, D., 1989
Exhibit at BENHS annual exhibition 1988. BENHS exhibition report. *Br. J. Ent. nat. Hist.* **2** (1): 46

Morris, R.K.A., ed., 1984
The Ecology of Mitcham Common. Unpublished report for the Board of Conservators of Mitcham Common.

Morris, R.K.A., 1991
Territorial behaviour in *Pipiza luteitarsis* and other Syrphidae. Hoverfly Newsletter No. 12.

Morris, R.K.A., 1993
Pipiza lugubris: a possible habitat link. Hoverfly Newsletter No. 16

Morris, R.K.A., & Collins, G.A., 1991
On the hibernation of Tissue moths *Triphosia dubitata* L. and the Herald moth *Scoliopteryx libatrix* L. in an old fort. *Entomologist's Rec. J. Var.* **103**: 313-321

Morton, A.J., & Collins, G.A., 1992
Distribution analysis of Surrey Lepidoptera using the DMAP computer package. *Nota lepid.* **15** (1): 84-88

Müller, A., 1872
Eristalis tenax attracted by painted flowers. *Entomologist's mon. Mag.* **8**: 273-274

Nixon, G.E., 1934
Two notes on the behaviour of *Volucella pellucens* in its association with the wasps *Vespa vulgaris* Linn. and *Vespa germanica* Fab. *Entomologist's mon. Mag.* **70**: 17-18

Parfit, R.W., 1947
SLENHS field meeting: Gomshall, 18th August 1946. *Proc. S. Lond. ent. nat. Hist. Soc.* **1946-47**: 75-76

Parmenter, L., 1935
Diptera on Bookham and Effingham Commons, Surrey. *Ent. Rec. J. Var.* **47**: 40

Parmenter, L., 1938
Some Diptera records. *London Naturalist* **17**: 75-76

Parmenter, L., 1941
Diptera visiting flowers of Devil's-bit Scabious, *Scabiosa succisa* L. *Entomologist's Rec. J. Var.* **53**: 134

Parmenter, L., 1942
The Diptera of Limpsfield Common. *London Naturalist* **21**: 18-33

Parmenter, L., 1949
Volucella zonaria (Poda) (Dipt, Syrphidae), some 1948 records with notes on the comparison with *V. inanis* F. *Entomologist's mon. Mag.* **85**: 38

Parmenter, L., 1950a
The Diptera of Bookham Common. *London Naturalist* **29**: 98-133

Parmenter, L., 1950b
Doros conopseus F. (Dipt., Syrphidae) in Surrey. *Entomologist's mon. Mag.* **86**: 256

Parmenter, L., 1951
The numbers of eggs laid by a hoverfly. *Entomologist's Rec. J. Var.* **63**: 255

Parmenter, L., 1952a
Flies at ivy bloom. *Entomologist's Rec. J. Var.* **64**: 90-91

Parmenter, L., 1952b

Further records of *Doros conopseus* F. (Dipt., Syrphidae) and *Myennis octopunctata* Coq. (Dipt., Otitidae) in Surrey. *Entomologist's mon. Mag.* **88**: 13

Parmenter, L., 1954

Syrphus euchromus Kowarz in Britain: its habits and habitat. *Entomologist's Rec. J. Var.* **66**: 122

Parmenter, L., 1955a

London Natural History Society Entomological section: July 19th 1955. *Entomologist's mon. Mag.* **91**: xxxix

Parmenter, L., 1955b

Flies visiting the bluebell, *Endymion non-scriptus* (L.) Garcke. *Entomologist's Rec. J. Var.* **67**: 89-91

Parmenter, L., 1956a

London Natural History Society Entomological section, August 28th 1956. *Entomologist's mon. Mag.* **92**: xl

Parmenter, L., 1956b

Flies and the selection of flowers they visit. *Entomologist's Rec. J. Var.* **68**: 242-243

Parmenter, L., 1957

Diptera of the chalk downs in May, about Old Coulsdon, Surrey. *Entomologist's Rec. J. Var.* **69**: 126-130

Parmenter, L., 1960

A further list of the Diptera of Bookham Common. *London Naturalist* **39**: 66-76

Parmenter, L., 1966

Some additions to the list of flies (Diptera) of Bookham Common. *London Naturalist* **45**: 56-59

Parmenter, L., 1968

Some records of Diptera predators and their prey. *Proc. Brit. ent. nat. Hist. Soc.* **1**: 37-42

Payne, R.M., 1949

London Natural History Society, September 28th, 1948. *Entomologist's mon. Mag.* **85**: 24

Perry, I., 1996

Callicera aurata in Suffolk found breeding in birch. *Dipterists Digest* (new series) **3** (2): 53

Plant, C.W., 1986
> A working list of hoverflies (Diptera: Syrphidae) of the
> London area. *London Naturalist* **65**: 109-117

Ranger, J.A., 1955
> *Volucella zonaria* Poda, in Surrey and Middlesex. *Amateur*
> *Entomologists' Society Bulletin* **14**: 24

Richardson, A.E., 1954
> SLENHS field meeting: Bookham 22nd June 1952. *Proc. S.*
> *Lond. ent. nat. Hist. Soc,* **1952/1953**. 81-82

Riley, J.A., 1946
> *Volucella zonaria* (Poda) (Dipt. Syrphidae) at Wimbledon.
> *The Entomologist* **79**: 247

Riley-Irving, J.E.C., 1947
> Observations. *Amateur Entomologists' Society Bulletin* **7**: 83

Rotheray, G.E., 1979
> *Atlas of the Diptera of Staffordshire part 1: Hoverflies.*
> Staffordshire Biological Recording Scheme publication No. 5,
> City Museum & Art Gallery, Stoke on Trent.

Rotheray, G.E., 1987
> The larvae and puparia of five species of aphidophagous
> Syrphidae (Diptera). *Entomologist's mon. Mag.* **123**: 121-125

Rotheray, G.E., 1993
> Colour guide to hoverfly larvae. *Dipterists Digest* **9**

Rotheray, G.E., 1996
> The larvae of *Brachyopa scutellaris* Robineau-Desvoidy
> (Diptera: Syrphidae), with a key to and notes on the larvae of
> British *Brachyopa* species. *Entomologist's Gazette*
> **47**: 199-205

Speight, M.C.D., 1973
> British species of *Sphaerophoria* (Diptera, Syrphidae)
> confused with *S. menthastri* (L.) Including a key to the males
> of seven species of *Sphaerophoria* found in the British Isles.
> *Entomologist* **106**: 228-233

Speight, M.C.D., 1988
> Syrphidae known from temperate Western Europe: potential
> additions to the fauna of Great Britain and Ireland and a
> provisional species list for N. France. *Dipterists Digest* **1**: 2-35

Spencer, J., 1986
Inventory of ancient woodlands: Greater London (provisional). Nature Conservancy Council, Peterborough. Unpublished.

Stace, C., 1991
New Flora of the British Isles Cambridge University Press, Cambridge.

Struthers, F.M., 1955
SLENHS field meeting: Colley Hill 22nd June 1953. *Proc. S. Lond. ent. nat. Hist. Soc.* **1953-54**. 82-83

Struthers, F.M., 1961
SLENHS field meeting: Horsley, Surrey 16th August 1959. *Proc. S. Lond. ent. nat. Hist. Soc.* **1959**. 89-90

Struthers, F.M., 1961
SLENHS field meeting: Bookham, Surrey 10th April 1960. *Proc. S. Lond. ent. nat. Hist. Soc.* **1960**. 75-76

Stubbs, A.E., 1967a
SLENHS field meeting: Sheepleas, Horsley, Surrey 21st May 1966. *Proc. S. Lond. ent. nat. Hist. Soc.* **1967**. 83-85

Stubbs, A.E., 1967b
SLENHS field meeting: Sheepleas, Horsley, Surrey 22nd April 1967. *Proc. S. Lond. ent. nat. Hist. Soc.* **1967**. 87

Stubbs, A.E., 1967c
Field meetings: Ockham, Surrey 20th August 1967. *Proc. S. Lond. ent. nat. Hist. Soc.* **1967**. 120-121

Stubbs, A.E., 1971a
BENHS field meeting: Gomshall, Surrey 30th May 1970. *Proc. Brit. ent. nat. Hist. Soc.* **4** (1): 23-24

Stubbs, A.E., 1971b
BENHS field meeting: Thursley, Surrey 6th September 1970. *Proc. Brit. ent. nat. Hist. Soc.* **4** (1): 27-28

Stubbs, A.E., 1971c
BENHS proceedings: 8th July 1971. *Proc. Brit. ent. nat. Hist. Soc.* **4** (3): 94-95

Stubbs, A.E., 1972
BENHS field meeting: Bookham Common, Surrey 15th August 1971. *Proc. Brit. ent. nat. Hist. Soc.* **4** (4): 126-127

Stubbs, A.E., 1980
Neocnemodon brevidens (Egger, 1865) (Diptera: Syrphidae)
new to Britain. *Entomologist's Rec. J. Var.* **92**: 45-46

Stubbs, A.E., 1981
Cyril Oswald Hammond. *Proc. Brit. ent. nat. Hist. Soc.*
14 (1/2): 40-43

Stubbs, A.E., 1982
Hoverflies as primary woodland indicators with reference to
Wharncliffe Wood. *Sorby Record* **20**: 62-67

Stubbs, A.E., 1991
A method of monitoring garden hoverflies. *Dipterists Digest*
10: 26-39

Stubbs, A.E., 1994
Sphegina (*Asiosphegina*) *sibirica* Stackelberg 1953, a new
species and sub-genus of hoverfly (Diptera, Syrphidae) in
Britain. *Dipterists Digest* (new series) **1** (1): 23-25

Stubbs, A.E., 1996
British Hoverflies (second supplement). British Entomological
and Natural History Society.

Stubbs, A.E., and Falk, S.J., 1983
British Hoverflies. British Entomological and Natural History
Society.

Uffen, R.W.J., 1958
SLENHS field meeting: Oxshott, Surrey 20th April 1957.
Proc. S. Lond. ent. nat. Hist. Soc. **1957**: 61-62

Uffen, R.W.J., 1961
SLENHS field meeting: Eashing, Surrey 4th June 1960. *Proc.
S. Lond. ent. nat. Hist. Soc.* **1960**: 82-83

Uffen, R.W.J., 1964
SLENHS field meeting: Gomshall, Surrey 1st June 1963. *Proc.
S. Lond. ent. nat. Hist. Soc.* **1963**: 66-67

Uffen, R.W.J., 1969
BENHS field meeting: Netley Heath, Surrey 31st August 1968.
Proc. Brit. ent. nat. Hist. Soc. **2** (1): 25-26

Vallins, F.T., 1956
SLENHS field meeting: Mickleham 29th August 1954. *Proc.
S. Lond. ent. nat. Hist. Soc.* **1954-55**: 92-93

Vallins, F.T., 1957
SLENHS field meeting: Cosford Mill, Thursley 21st August 1955. *Proc. S. Lond. ent. nat. Hist. Soc.* **1955**: 83-85

Verrall, G.H., 1901
British Flies: Platypezidae, Pipunculidae and Syrphidae of Great Britain (re-printed 1969). E.W. Classey

Wakeley, S., 1949
SLENHS field meeting: Ranmore (White Downs) 12th July 1947. *Proc. S. Lond. ent. nat. Hist. Soc.* **1947-48**: 65-67

Wakely, S., 1950
Doros conopseus F. Entomologist's Rec. J. Var. **62**: 97

Wakely, S., 1953
Volucella zonaria Poda at Camberwell. *Entomologist's Rec. J. Var.* **65**: 31

Wakely, S., 1955
Microdon devius L. Entomologist's Rec. J. Var. **67**: 90-91

Wakely, S., 1957
SLENHS field meeting: Eashing Moors 11th June 1955. *Proc. S. Lond. ent. nat. Hist. Soc.* **1955**: 74-75

Whicher, L.S., 1948
Volucella zonaria Poda (Dipt. Syrphidae) in Surrey. *Entomologist's mon. Mag.* **84**: 263

Whiteley, D., 1987
Hoverflies of the Sheffield area and North Derbyshire. *Sorby Record Special Series* No. **6**

Withers, P., 1983
Recent records of some rare British Syrphidae. *Entomologist's mon. Mag.* **119**: 11-12

Woodcock, A.J.A., 1956
Volucella zonaria Poda (Dipt. Syrphidae) in Kew Gardens, Surrey. *Entomologist's mon. Mag.* **92**: 258

APPENDIX 7 – Glossary

aculeate: having a sharp point – the aculeate Hymenoptera are the stinging bees and wasps.

antennal pits: pits located on the third segment of the antenna which have a sensory function. Some species have a single large and distinct pit, e.g. *Brachyopa*, but there are small scattered pits in *Cheilosia*.

aphidophagous: feeding on aphids.

arboreal: associated with trees.

arista: a long bristle arising from the third antennal segment

autecology: the biology of a single species and the environmental factors which affect it.

avermectins: chemical control of insect parasites of cattle. These chemicals are residual in the dung and limit the growth of invertebrate larvae which feed on it, resulting in much reduced levels of dung-feeding larvae and structural abnormalities in resulting adults.

biodiversity: the entire range of plants and animals. In order to save the species at risk of extinction in Britain, the UK Biodiversity Action Plan, published in 1994, provided the framework for this effort. It was followed up by the publication of three lists of species identified as in need of special action.

calypterate: a specialised group of flies with modified lobes at the base of the wings which cover the halteres. These flies are mainly extremely bristly or hairy.

chitinous: made of chitin which forms the hard exoskeleton of an insect; each plate of the exoskeleton is held together by a more flexible membrane which is exposed in places such as the underside at the juncture of the thorax and abdomen.

composite (flower): a member of the family Compositae, characterised by compound flower heads comprising a number of individual florets.

coxa(e): the basal joint of the leg, linking it with the thorax.

dimorphism: differences in the size or colouration between broods of the same species, usually associated with the conditions under which they develop; those which develop in colder conditions are often smaller or darker.

double-brooded: having two broods or generations in a single year.

ectoparasite: a parasite which lives on the exterior of an organism.

gravid: carrying eggs – often the abdomen appears to be swollen.

gynandromorph: an individual exhibiting a mosaic of male and female characters.

halteres: the modified second pair of wings which act as balancing organs.

humeri: swellings at the front of the thorax which form the "shoulders"; they are often hidden by the head, but are exposed in the Milesiinae and Microdontinae.

in copula: a pair joined in copulation.

lunulate: shaped like a small crescent.

melanic: a dark or black form of an otherwise patterned species.

Nationally Scarce: species known from fewer than 100 and more than 15 10 km squares across Great Britain.

ovipositing: laying eggs.

pH: a measure of the acidity of solutions, ranging from 1 to 14; neutral solutions are pH7; acidic solutions are less than pH7 and alkaline solutions are greater than pH7.

phenology: the timing of emergence in relation to climatic and other factors.

phylogeny: the study of the history of development of a species, represented in the arrangement of species as tribes, genera and sub-genera.

phytophagous: plant-feeding.

produced (as in face): extended forwards beyond the normal shape for the genus.

puparium: the last larval skin which forms a casing of the pupal stage when the insect changes from a larva to an adult.

Red data book: the listing of species known or thought to be under some degree of threat in Great Britain. There are three key categories:

    RDB1   Endangered
    RDB2   Vulnerable
    RDB3   Rare (known from 15 or fewer 10km squares across Great Britain)

saproxylic: describes species whose larval stages develop in dead wood.

scutellar: referring to a shield-shaped structure at the rear of the thorax.

supra generic: above the level of genus.

tarsal: of the foot; in flies the tarsi comprise 5 segments, the last bearing claws and sometimes pads (pulvilli). In *Platycheirus* the metatarsus or basal segment is sometimes modified to a considerable degree and bears pits which are highly distinctive.

tergites: the upper abdominal plates of a fly.

tetrad: a two kilometre square.

thorax: the main body which bears the legs.

tibia(e): the fourth joint of the leg, which is normally one of just two elongate sections (femur and tibia).

INDEX

Figures in bold indicate plate numbers

Anasimyia 147
    contracta 147,**13**
    lineata 148
    lunulata 148
    transfuga 148
Baccha 51
    elongata 52
    obscuripennis 52
Brachyopa 131
    bicolor.................. 131
    insensilis 131,**7**
    pilosa 133
    scutellaris........... 134
Brachypalpoides 179
    lentus................... 179
Brachypalpus 180
    laphriformis 180,**6**
Caliprobola 181
    speciosa 181,**4**
Callicera 108
    aurata............... 108,**5**
    (aenea) 108
Chalcosyrphus 182
    nemorum 182
Cheilosia 109
    albipila........... 109,**15**
    albitarsis 111
    antiqua 111
    barbata 112
    bergenstammi 113
    carbonaria 113
    cynocephala 114
    fraterna 114
    griseiventris 115
    grossa 115,**15**
    (honesta) 117
    illustrata 116,**1**
    impressa 116
    (intonsa) 117
    lasiopa 117
    latifrons............... 117
    longula 118
    mutabilis 118
    (nasutula) 125

nebulosa.............. 119
nigripes 119
pagana 120
praecox 120
proxima 121
scutellata............. 121
semifasciata 122
soror 122
variabilis 124
velutina 125
vernalis 125
vicina 125
vulpina 126
Chrysogaster 135
    cemiteriorum 135
    (chalbeata) 135
    (hirtella) 138
    (macquarti) 137
    solstitialis........... 135
    virescens............. 136
Chrysotoxum 65
    arcuatum 65
    bicinctum 66
    cautum 66,**9**
    elegans 67
    festivum 67,**9**
    octomaculatum...... 68
    verralli 69
Criorhina 182
    asilica 182,**4**
    berberina 183
    floccosa............ 184,**4**
    ranunculi............. 184
Dasysyrphus.............. 71
    albostriatus 71
    friuliensis 71
    (lunatus) 72
    pinastri.................. 72
    tricinctus 72,**3**
    venustus 73
Didea 73
    fasciata 73,**3**
    intermedia 75

Doros 75
    (conopseus) 75
    profuges 75,**8**
Epistrophe 76
    diaphana 76
    eligans............... 78,**2**
    grossulariae 78,**1**
    melanostoma 79
    nitidicollis 80
(Epistrophella) 91
    (euchroma) 91
Episyrphus 82
    balteatus 82
Eriozona 83
    erratica 83
Eristalis 150
    abusivus 150
    arbustorum........... 150
    horticola 151
    intricarius 152
    interruptus 151,**15**
    (nemorum).......... 151
    pertinax 152,**1**
    tenax 153
Eristalinus 149
    aeneus 149
    sepulchralis 149
Eumerus 159
    ornatus 159
    strigatus 160
    tuberculatus 160
Eupeodes 83
    corollae 83
    latifasciatus........... 84
    latilunulatus 84
    luniger.................. 85
    nitens.................... 85
Ferdinandea 126
    cuprea 126
    ruficornis 127
Helophilus................. 154
    hybridus 154,**13**
    pendulus 154
    trivittatus............. 155

INDEX

Figures in bold indicate plate numbers

Heringia 163
   brevidens 164
   heringi 163
   latitarsis 164
   pubescens 165
   vitripennis 165

Lejogaster 136
   metallina 136
   tarsata 137
   (splendida) 137

Leucozona 86
   glaucia 86,**1**
   laternaria 86
   lucorum 87,**2**

Mallota 155
   cimbiciformis 155

(Megasyrphus
   annulipes) 83

Melangyna 87
   barbifrons 87
   cincta 88
   compositarum 88
   labiatarum 89,**2**
   lasiophthalma 89
   quadrimaculata 90
   umbellatarum 90

Melanogaster 137
   aerosa 137
   hirtella 138

Melanostoma 53
   mellinum 53
   scalare 53

Meligramma 91
   euchromum 91,**15**
   guttatum 92
   trianguliferum 92

Meliscaeva 93
   auricollis 93
   cinctella 94

Merodon 161
   equestris 161,**14**

(Metasyrphus) 83

Microdon 192
   analis 192,**11**
   devius 193,**8**
   (eggeri) 192
   mutabilis 194

Myathropa 156
   florea 156,**6**

Myolepta 138
   dubia 138
   (luteola) 138

Neoascia 139,**12**
   geniculata 140
   interrupta 141
   meticulosa 141
   obliqua 141
   podagrica 139
   tenur 140

(Neocnemodon) 163

Orthonevra 142
   brevicornis 142
   geniculata 143
   nobilis 143
   (splendens) 144

Paragus 64
   albifrons 65
   haemorrhous 64
   tibialis 64

Parasyrphus 94
   annulatus 94
   lineola 95
   malinellus 95
   punctulatus 96
   vittiger 96

Parhelophilus 157,**13**
   frutetorum 157
   versicolor 158

Pelecocera 162
   tricincta 162

Pipiza 166
   austriaca 166
   bimaculata 166
   fenestrata 167
   lugubris 167
   luteitarsis 168
   noctiluca 168

Pipizella 170
   (varipes) 171
   viduata 171
   virens 171
   maculipennis 170

Platycheirus 54
   albimanus 55
   ambiguus 54
   angustatus 55
   clypeatus 56
   discimanus 57
   fulviventris 58
   granditarsus 62
   immarginatus 58
   manicatus 58,**8**
   occultus 59
   peltatus 59
   rosarum 63
   scambus 60
   scutatus 61
   sticticus 61
   tarsalis 62

Pocota 185
   personata 185,**5**

Portevinia 128
   maculata 128,**2**

Psilota 161
   anthracina 161

Rhingia 129
   campestris 129
   rostrata 130

Riponnensia 144
   splendens 144

Scaeva 97
   pyrastri 97
   selenitica 97

INDEX

Figures in bold indicate plate numbers

Sericomyia 174
   lappona 174,**10**
   silentis 174,**10**

Sphaerophoria 98
   (abbreviata) 99
   batava 98
   fatarum 99
   interrupta 99
   (menthastri) 99
   philanthus 100
   rueppellii 100
   scripta 101
   taeniata 101
   virgata 102

Sphegina 145
   clunipes 145
   elegans 146
   (kimakowiczi) 146
   verecunda 146

Syritta 186
   pipiens 186

Syrphus 105
   ribesii 105,**2**
   torvus 105
   vitripennis 106

Trichopsomyia 172
   flavitarsis 172

Triglyphus 173
   primus 173

Tropidia 187
   scita 187,**13**

Volucella 175
   bombylans 175,**16**
   inanis 175,**16**
   inflata 176
   pellucens 177,**16**
   zonaria 177,**16**

Xanthandrus 63
   comtus 63

Xanthogramma 107
   citrofasciatum .. 107,**9**
   pedissequum 107,**9**

Xylota 188
   abiens 188
   florum 189
   segnis 189
   sylvarum 190,**4**
   tarda 190
   xanthocnema 191